Uncommon Mathematical Excursions

Polynomia and Related Realms

Figure 2.7 (p. 35). Courtesy Niloufer Mackey.

Photo in Sidebar 4.2 (p. 81). "I Have a Photographic Memory", 1987, Bloomington to Santa Barbara and back, 1975–1978, Paul Halmos, photograph number 470, Copyright © 1987 by the American Mathematical Society. All rights reserved.

Photo in Sidebar 6.1 (p. 123). Courtesy Architect of the Capitol.

Figures 7.9 and 7.10 (pp. 148, 149). Courtesy Rob Ives, http://www.flying-pig.co.uk.

Photo in Sidebar 8.2 (p. 168). Courtesy Gerald J. Porter.

Cartoon p. 173. Courtesy John de Pillis.

Figures 12.1 and 12.2 (pp. 234, 235). Courtesy NASA/JPL-Caltech.

© 2009 by
The Mathematical Association of America (Incorporated)
Library of Congress Catalog Card Number 2008940929
ISBN 978-0-88385-341-2
Printed in the United States of America
Current Printing (last digit):
10 9 8 7 6 5 4 3 2 1

The Dolciani Mathematical Expositions

NUMBER THIRTY-FIVE

Uncommon Mathematical Excursions

Polynomia and Related Realms

Dan Kalman
American University

Published and Distributed by
The Mathematical Association of America

The DOLCIANI MATHEMATICAL EXPOSITIONS series of the Mathematical Association of America was established through a generous gift to the Association from Mary P. Dolciani, Professor of Mathematics at Hunter College of the City University of New York. In making the gift, Professor Dolciani, herself an exceptionally talented and successful expositor of mathematics, had the purpose of furthering the ideal of excellence in mathematical exposition.

The Association, for its part, was delighted to accept the gracious gesture initiating the revolving fund for this series from one who has served the Association with distinction, both as a member of the Committee on Publications and as a member of the Board of Governors. It was with genuine pleasure that the Board chose to name the series in her honor.

The books in the series are selected for their lucid expository style and stimulating mathematical content. Typically, they contain an ample supply of exercises, many with accompanying solutions. They are intended to be sufficiently elementary for the undergraduate and even the mathematically inclined high-school student to understand and enjoy, but also to be interesting and sometimes challenging to the more advanced mathematician.

26. *Mathematical Diamonds,* Ross Honsberger
27. *Proofs that Really Count: The Art of Combinatorial Proof,* Arthur T. Benjamin and Jennifer J. Quinn
28. *Mathematical Delights,* Ross Honsberger
29. *Conics,* Keith Kendig
30. *Hesiod's Anvil: falling and spinning through heaven and earth,* Andrew J. Simoson
31. *A Garden of Integrals,* Frank E. Burk
32. *A Guide to Complex Variables* (MAA Guides #1), Steven G. Krantz
33. *Sink or Float? Thought Problems in Math and Physics*, Keith Kendig
34. *Biscuits of Number Theory*, Arthur T. Benjamin and Ezra Brown
35. *Uncommon Mathematical Excursions: Polynomia and Related Realms*, Dan Kalman

MAA Service Center
P.O. Box 91112
Washington, DC 20090-1112
1-800-331-1MAA FAX: 1-301-206-9789

For Linda,
my first and best collaborator.

Preface

This book is a travel guide to three mathematical realms. It is not for beginners. The intended audience includes those who are already very familiar with the most popular attractions of the three realms. So why do they need a travel guide at all?

By way of analogy, think about tourists who travel to France. Many will visit France just once, or perhaps a few times. On these visits they will take in the most popular and famous sights and destinations: the Eiffel Tower, the Louvre, and so forth. But there are others who visit France many times, forging a special fondness for the place and its people. These travelers have been to all the most popular attractions, and now wish to expand their French experience by seeking out lesser known destinations.

It is much the same with mathematics. Most who study beyond the minimum demanded by high school or college follow a common path through a standard curriculum: algebra, precalculus, calculus, and perhaps as far as linear algebra and differential equations. These students are like those who visit France just once or twice. In their limited exposure to the subject, they focus on the main attractions. This book is not for them, or at least, not primarily for them.

Rather, it is for their teachers. The teachers have traveled through the standard curriculum repeatedly, until it is thoroughly comfortable and familiar. When it comes to touring these mathematical realms, they are old hands, indeed. And like those many-time visitors to France, they may wish to seek out new vistas and explore unsuspected wonders a bit off the beaten track.

It is also for anyone who has a long history of applying mathematics, including scientists, engineers, and analysts. They, too, are intimately acquainted with the core ideas of algebra, calculus, and related subjects. They, too, might be interested in seeing some new aspects of an old familiar terrain.

The mathematical subjects that contain the standard curriculum are incredibly rich. There are countless extensions, digressions, and supplementary topics that illuminate and enrich the core. Moreover, these topics are highly accessible to the seasoned mathematical hands who have journeyed so many times through these realms. Anyone who has taught precalculus a few times will be well prepared to explore additional properties of polynomials, rational functions, and exponentials, for example.

In fact, there are so many possible journeys into the mathematical back country, as it were, that it would be impossible to survey even a small percentage of them in a single volume. This book makes no pretense of doing so. Rather, I have gathered together some topics that are particularly appealing and familiar to me, in the hope that these topics will be interesting and appealing to others, as well.

The title and organization of the book follow the metaphor introduced above. The material is presented as a series of explorations in three mathematical realms, focussing on polynomials, max/min problems, and calculus. A thorough familiarity with the standard curriculum should be sufficient background for all of the excursions. However, for those who wish to brush up on parts of this material, or who might be accustomed to different notations or terminology than I use in the main text, I have prepared some appendices. These are available at a companion website for this book, see [87]. There the reader will also find some additional supplementary material, including animated and interactive graphics displays for some of the topics discussed in the book.

Although this book was written with teachers in mind, it is really for anyone who holds a deep affection for the core curriculum, and wishes to study more of it. This can include specialists in allied fields who retain a strong mathematical inclination, as well as students who are particularly attracted to the subject. They will find curious connections, surprising patterns, and unexpected insights about old and familiar ideas.

So pack your bags, get a stout walking stick, and meet me at the portal on page 1. Wonders await!

Acknowledgments

The first draft of this book was written during a sabbatical leave from American University. I am grateful for the university's support of this project, and for providing me an academic and professional home for the past fifteen years.

The initial idea for this book grew out of discussions with Maurice Burke. He gave valuable advice and suggestions on the first several chapters, and was a strong source of encouragement for the project.

Underwood Dudley also offered constant encouragement and support. In addition, he painstakingly read preliminary drafts of all the chapters. His many suggestions for improvement had a great impact on the project. Woody, to quote a wise man, "The editor is always right."

Don Albers also provided early feedback and strong encouragement and support. Thanks again, Don, for your confidence in this project, and those that came before it.

Bruce Torrence showed me how to generate some beautiful three-dimensional diagrams using mathematical software. As in so many other instances, his enthusiasm and energy were much appreciated.

Doug Ensley developed a beautiful interactive presentation of envelopes of families of curves, now available at UME's companion website. Doug created it using the tools he has been developing (with his coauthor, Barbara Kaskosz) for hosting mathematical explorations with Flash software. This activity is one of the highlights of the UME website.

I am grateful to John de Pillis for contributing original artwork to this volume. This was an especially gracious act of generosity, given the fact that his cartoon illustrates a pun that might charitably be described as 'lame.'

A number of other colleagues and correspondents deserve special thanks. Lowell Beineke supplied me with references for the *perplexing polynomial puzzle*. Dave Renfro and David Singmaster provided helpful references on the ladder problem. Ed Barbeau, Craig Fraser, and Bruce Pourciau helped me find appropriate references for Lagrange multipliers, and Bruce gave valuable suggestions about the *Leveling with Lagrange* chapter. Jennifer Kalman Beal translated some of Lagrange's work for me. Douglas Rogers and Pat Ballew provided information about Lill's method, and about Lill himself. Douglas was especially generous in tracking down and translating an Austrian bibliographic citation. Frederick Rickey and Rüdiger Thiele also helped with that translation. And Fred generously helped me track down parts of the Lill story that intersect with the US Military Academy, and with the Artemis Martin collection at my home institution, American University. Mike McNamee gave me information about a couple of Lill's papers. Harold M. Edwards helped me understand the history of the fundamental theorem of symmetric functions.

Several times I sought assistance by posting queries to the Project NExT discussion list, and I was always rewarded with several prompt responses. Without mentioning by name all of those who corresponded with me, let me express my gratitude. Indeed, I am grateful to be a member of so warm, friendly, and helpful a community of mathematicians and scholars.

Thanks, too, are due to Elaine Pedreira Sullivan, Associate Director for Publications, and Beverly J. Ruedi, Electronic Production Manager, at the MAA. Their efforts are indispensable in transforming manuscripts into books, and they deserve full credit for the attractive appearance of this volume.

Finally, I would like to acknowledge all the neighbors, personal friends, and family members who encouraged me throughout the long effort that produced this book. Especially I want to thank the members of the Hatrack River Writers Workshop group I have been a part of since 2004. As aspiring and successful authors, they know what is involved in writing a book. I am so grateful for their sympathetic support and steadfast encouragement.

Contents

Part I

The Province of Polynomia

N journeying through the elementary mathematics curriculum, who has not been struck by the sublime beauty of Polynomia's countryside? From her rolling hillsides to her gently rounded valleys, the landforms undulate to rhythms that are hypnotic and infinitely smooth. This is a land steeped in mythic history, where great deeds have been done and great quests pursued. Here the foundations were laid for hallowed traditions extending throughout all of mathematics. As all mathematicians know, if you want to find your roots, begin in Polynomia!

Travelers in the world of mathematics make frequent visits to Polynomia, beginning very near the start of the study of algebra. This is because polynomials are among the simplest algebraic objects, and at the same time are among the most useful. They have many interesting and useful properties that make them a natural choice for applications and continued study through the introductory curriculum, and on into much more advanced levels.

Why are polynomials so desirable? For one thing, they avoid the dreaded operation of division. Algebra is an extension of arithmetic, after all, and in arithmetic what trouble there is comes from division. Any algebraic expression, such as $4x - 5$, is really a recipe for calculation. Give me a value for x and the recipe directs me to compute a related quantity. Polynomials encompass all of the recipes that can be performed without division. Anyone who much prefers addition, subtraction, and multiplication to division will understand the attraction of working with polynomials.

A far stronger attraction of polynomials comes from their many interesting properties. In elementary mathematics, a host of these properties are standard fare: aspects of their graphs, patterns that arise in factoring, the simple rules for differentiation and integration, to name a few. These are the precincts most popular in the mainstream of Polynomia tourism. But there are additional destinations that are just as interesting, maybe even more interesting, just a little off the beaten path. Some were popular attractions years ago, when tours of analytic geometry and theory of equations were highly fashionable. That was when calculus was an advanced topic and the pre-calculus curriculum had greater depth. There are also other aspects of polynomials, of a more modern flavor, that are rarely encountered in standard curricula.

This first part of the book is dedicated to the picturesque province of Polynomia. It will guide the seasoned mathematical traveler to interesting aspects of polynomials that are easily within reach from the more popular fare of the standard curricula. These provide a rich resource for extending and deepening the understanding and appreciation of the polynomials that are so familiar in elementary mathematics.

In the next several chapters, you will find discussions of

- An alternative to the standard descending and factored forms;

- A little known geometric construction of polynomial roots;

- Connections between roots and coefficients;

- The consequence of reversing the coefficients;

- Fitting a polynomial curve to a set of points;

- Extensions of the quadratic formula to cubic and quartic equations;

as well as connections between these topics and to other parts of mathematics.

Long Division of Polynomials

Here is an example of polynomial long division. The object is to divide $x^4 + 5x^3 - 3x^2 + 2x + 1$ by $x^2 + 2x - 1$. Using the same format as whole number long division produces the following results.

$$
\begin{array}{r}
x^2 + 3x - 8 \\
x^2 + 2x - 1 \overline{\smash{\big)}\ x^4 + 5x^3 - 3x^2 + 2x + 1} \\
x^4 + 2x^3 - x^2 \\
\hline
3x^3 - 2x^2 + 2x \\
3x^3 + 6x^2 - 3x \\
\hline
-8x^2 + 5x + 1 \\
-8x^2 - 16x + 8 \\
\hline
21x - 7
\end{array}
$$

The top line of the computation gives the quotient, $x^2 + 3x - 8$. The bottom line of the computation shows the remainder $21x - 7$. All four polynomials, the divisor, divisee, quotient, and remainder are linked in the equation

$$x^4 + 5x^3 - 3x^2 + 2x + 1 = (x^2 + 2x - 1)(x^2 + 3x - 8) + 21x - 7.$$

Before embarking on these explorations, we should agree upon some terminology and notation. Any polynomial in a single variable t can be expressed in *descending form*

$$a_n t^n + a_{n-1} t^{n-1} + \cdots + a_1 t + a_0.$$

The numbers a_0 through a_n are called the *coefficients*. We generally assume a_n is not zero, and when that is true, n is called the *degree* of the polynomial. Also when a_n is nonzero, it is referred to as the *leading coefficient*. A *monic polynomial* is one with a leading coefficient of 1. So, for example, $x^3 - 4x^2 + 2x - 7$ is a monic polynomial of degree 3. Polynomials of degree 1 are called *linear*, those of degree 2 are called *quadratic*, and so on for higher degrees: *cubic* for degree 3, *quartic* for degree 4, and *quintic* for degree 5.

Polynomials will often be denoted by a single letter: e.g. p or q. On the other hand polynomials are functions and so may be denoted using function notation, as in

$$p(x) = x^2 + 2x - 3.$$

Values of x for which $p(x) = 0$ have particular significance. We call these *roots* or *zeros* of the polynomial. For the example above, -3 is a root of p because $p(-3) = 0$. More generally, a solution to a polynomial equation will be called a root or zero of the equation.

Any constant is a polynomial. If the constant is nonzero, then it is a polynomial of degree 0. If it is the constant zero, however, there is no leading coefficient and there is no defined degree. This exceptional specimen is called the *zero polynomial*.

Sometimes it is useful to make assumptions about the coefficients of a polynomial, for example, that they are all integers, or all rational numbers, or all positive. Unless otherwise stated, we will assume the coefficients are real numbers.

As mentioned polynomials have the virtue that they can be formulated without mention of division. Paradoxically, we often need to divide one polynomial by another. In fact, the process for carrying out polynomial division is modeled on the long division algorithm for integers, as illustrated in Sidebar I.1. Polynomials can always be divided in like fashion, either working out evenly, or producing a non-zero remainder. A general statement of this principle follows.

Polynomial Division. Given polynomials f and g, with f nonzero, division of g by f produces a quotient polynomial q and a remainder polynomial r, where either r is the zero polynomial, or r has degree less than the degree of f. In the former case, $g = fq$, and we say that f is a *divisor* or a *factor* of g. In the latter case, $g = fq + r$.

Division of polynomials leads to the following useful consequences.

1. **The Factor Theorem**: a is a root of polynomial $p(x)$ if and only if $(x - a)$ is a factor of $p(x)$.

2. **The Remainder Theorem**: Division of a polynomial $p(x)$ by $x - a$ results in a remainder equal to $p(a)$.

3. **The Root-count Theorem**: A polynomial of degree n can have at most n distinct roots.

A more detailed account of polynomial division and its consequences is available at the website for this book [87].

1

Horner's Form

The most familiar representations of a polynomial are descending form (as described earlier) and *factored form*. For example, the equation

$$x^4 + x^3 - x^2 - x = x(x-1)(x+1)^2$$

shows both descending and factored forms for one polynomial. We also know that any polynomial can be expressed in other forms by rearranging the order of the terms, grouping, or a variety of other manipulations. There is an additional *standard* form, less well known than descending and factored forms, that is worth knowing about. It is *Horner's form*.

The idea of Horner's form can be seen in an example. If

$$p(x) = 5x^4 - 11x^3 + 6x^2 + 7x - 3, \tag{1}$$

its Horner's form is

$$p(x) = (((5x - 11)x + 6)x + 7)x - 3. \tag{2}$$

Both forms have the same sequence of coefficients, 5, −11, 6, 7, −3. However, in Horner's form there are no exponents, because of the grouping. In fact, one way to derive Horner's form from the descending form is successive grouping and division by x. To see this, look first at the descending form (1). Factor x out of all the terms except the constant, giving

$$p(x) = (5x^3 - 11x^2 + 6x + 7)x - 3.$$

Now apply the same process repeatedly, each time working on the part in the innermost set of parentheses. After four iterations, Horner's form is obtained.

Horner's form provides an efficient way to compute the value of a polynomial. Continuing with the same example, suppose you would like to compute $p(2)$. Using the Horner form (2), work from the innermost parenthesis outward, accumulating the result as you go along:

Operation	Result
Start with	5
times 2	10
minus 11	−1
times 2	−2
plus 6	4
times 2	8
plus 7	15
times 2	30
minus 3	27.

Thus $p(2) = 27$. With a small amount of practice, this is a feasible method for mentally computing values of polynomials. For this example, I can compute $p(2/3)$ in my head, but just barely. Can you find $p(3/5)$?

Even using a calculator, Horner's method is a time saver. Compare the key sequences needed for the descending form and the Horner form, assuming the calculator has a power key ($\wedge$), a square key (x^2), and an enter key ($=$):

Descending Form: $5 \times 2 \wedge 4 - 11 \times 2 \wedge 3 + 6 \times 2\ x^2 + 7 \times 2 - 3 =$

Horner Form: $5 \times 2 - 11 = \times 2 + 6 = \times 2 + 7 = \times 2 - 3 =$

The first method requires 24 keystrokes, while the second takes only 22. This difference might not seem like a great deal, but it can provide a competitive edge in calculator races. (You *do* race calculators, don't you?)

More seriously, the extra efficiency of Horner's method makes it attractive in computer programs, for situations requiring large numbers of polynomial evaluations. Though it may not be obvious, Horner's form is also less subject to roundoff error. We will return later to the topic of computational advantages. For now, we turn our attention to another aspect of Horner's form.

1.1   Horner's Form and Synthetic Division

In calculating with Horner's form, there is a simple pattern of operations: add a coefficient and multiply by x, add a coefficient and multiply by x, over and over. This suggests laying things out in a table for hand calculation. Here is one possible arrangement.

$$
\begin{array}{r|rrrrr}
2 & 5 & -11 & 6 & 7 & -3 \\
  &   & 10 & -2 & 8 & 30 \\
\hline
  & 5 & -1 & 4 & 15 & 27
\end{array}
$$

The left-most 2 on the first line, called the *boxed entry*, is the value of x. It is used repeatedly as a multiplier. The remaining entries on the first line are the coefficients of the polynomial. Initially, the second and third lines are blank, except that the leading coefficient (5 in this case) is entered on the third line directly below its position on the first line. Now the computational process proceeds as follows.

- Multiply the right-most entry on the third line by the boxed entry.

- Enter the result one position to the right in the second line.

- Add this newest entry to the one above it, and enter the result below the line.

- Repeat.

Does this look familiar? It should — it is exactly the same as the tableau for synthetic division of the original polynomial, $p(x)$, by the linear factor $x - 2$. In general, evaluating a polynomial $p(x)$ at $x = a$ using the Horner form is identical to synthetic division of $p(x)$ by $x - a$.

This observation is closely related to the Remainder Theorem (page 5). The Remainder Theorem says that the remainder after dividing $p(x)$ by $x - a$ is the same as the value of $p(a)$. We can also see this by identifying synthetic division with the Horner form evaluation of $p(x)$. The final entry on the third line is then seen to be both the remainder from the division and the value for $p(a)$.

So evaluating a polynomial in Horner's form is actually synthetic division. This tells us something more about Horner evaluation. Think of that evaluation as a series of multiply-add steps. At the end of each step, we reach an intermediate stage of the computation. (These are exactly the intermediate results that appear each time the $=$ key is pressed in the calculator implementation on page 8.) But in the synthetic division interpretation these are the coefficients for the quotient $p(x)/(x - a)$. Therefore, the intermediate results encountered in Horner evaluation give the coefficients of the quotient.

Interestingly, each of the intermediate steps in Horner evaluation can be viewed as another polynomial in Horner form. For the example $p(x) = 5x^4 - 11x^3 + 6x^2 + 7x - 3$, these intermediate polynomials in both Horner form and descending form are shown below.

Horner Form	Descending Form
5	5
$5x - 11$	$5x - 11$
$(5x - 11)x + 6$	$5x^2 - 11x + 6$
$((5x - 11)x + 6)x + 7$	$5x^3 - 11x^2 + 6x + 7$
$(((5x - 11)x + 6)x + 7)x - 3$	$5x^4 - 11x^3 + 6x^2 + 7x - 3$

Although the polynomials are expressed in terms of the variable x, it actually makes more sense to use a different variable. Remember that they represent intermediate results when $p(x)$ is evaluated for a particular x. If we think of performing the calculation with $x = a$, then each of the intermediate results should be expressed in terms of a, rather than x. Moreover, when the intermediate results are considered to be coefficients in the quotient polynomial, they each are multiplied by a power of x. This, too, suggests expressing the intermediate results in terms of a.

To better examine the nature of the coefficients in the quotient, let us modify the preceding table. As before, we'll include the intermediate results from the Horner evaluation, but this time in terms of a, and only in descending form. The final row we can omit, since that pertains only to the remainder, and here we are interested in the coefficients for the quotient. Next, let us add a column showing the appropriate power of x for each of these coefficients. Finally, include a third column showing each coefficient multiplied by its power of x. Here is the result.

Coefficient	Power of x	Product
5	x^3	$5x^3$
$5a - 11$	x^2	$(5a - 11)x^2$
$5a^2 - 11a + 6$	x	$(5a^2 - 11a + 6)x$
$5a^3 - 11a^2 + 6a + 7$	1	$(5a^3 - 11a^2 + 6a + 7)$

Adding the entries in the last column, we obtain the quotient $p(x)/(x-a)$, disregarding the remainder r. In equation form,

$$\frac{p(x)}{x-a} = 5x^3 + (5a - 11)x^2 + (5a^2 - 11a + 6)x + (5a^3 - 11a^2 + 6a + 7) + r.$$

Here it is clear why we need two variables. The a represents a constant, and defines the divisor $x - a$. The x is the polynomial variable for divisor, divisee, and quotient. Using the same letter for both roles doesn't make sense.

But let's do it anyway! The next table is just like the previous one, except that all of the a's have been changed back to x's.

Coefficient	Power of x	Product
5	x^3	$5x^3$
$5x - 11$	x^2	$5x^3 - 11x^2$
$5x^2 - 11x + 6$	x	$5x^3 - 11x^2 + 6x$
$5x^3 - 11x^2 + 6x + 7$	1	$5x^3 - 11x^2 + 6x + 7$

Now something interesting happens if we again add the results in the final column. Do you see it? If not, look at the pattern in the final column. Where in the original polynomial we had one term $5x^4$, the last column of the table has *four* $5x^3$ terms. Where the original polynomial had one $-11x^3$ term, we now have *three* $-11x^2$ terms. And it is apparent that the same thing will happen for any polynomial: if the original polynomial has a term bx^k, the table will have k terms of bx^{k-1}. Adding up all of these entries produces kbx^{k-1}. Or, in a table like the one above, adding up the entries in the final column produces the derivative $p'(x)$ of the original polynomial.

Why is that? The answer has two parts. First, it is significant that we are ignoring the remainder. In general, when we divide $p(x)$ by $(x - a)$, we obtain a quotient $q(x)$ and a remainder r, where

$$p(x) = (x - a)q(x) + r.$$

But we know the remainder is really $p(a)$, so the equation can be written

$$p(x) = (x - a)q(x) + p(a).$$

This shows that $p(x) - p(a)$ is always evenly divisible by $x - a$, and that the quotient we find with synthetic division is $q(x) = (p(x) - p(a))/(x - a)$.

Do you sense the derivative creeping in? That expression, $(p(x) - p(a))/(x - a)$, is a *difference quotient*. It tells the slope of a line between two points on the graph of p, one at x and the other at a. That brings us to the second part of the answer. By expressing the coefficients in terms of x rather than a, we are in essence forming the difference quotient $(p(x) - p(x))/(x - x)$. Normally, this would be the indeterminate $0/0$. But here the difference quotient $(p(x) - p(a))/(x - a)$ is a polynomial, so substituting x for a is identical to taking a limit as a approaches x. We can almost dispense with the concept of limit entirely. As Newton did, we can assume a is different from x for the purposes of forming $(p(x) - p(a))/(x - a)$, and then set a equal to x to find the derivative. The table combines both steps into a single operation.

Using this analysis, we can derive the usual rules for derivatives of polynomials. This doesn't really need Horner's form. An understanding of synthetic division suffices. But Horner's form and synthetic division are so closely connected, it is hard to talk about one without mentioning the other. In any case, similar ideas to those just considered can be used to develop differentiation in a way that more or less avoids the limit concept. This will be explored in greater detail in Chapter 11. For now, the story of Horner's form continues.

1.2 Horner's Form and Efficient Computation

Years ago, when computers were much slower, the increased efficiency of Horner's form polynomial evaluation made it a standard tool in scientific programming. Today, this is not so great a consideration for computing $p(x)$ when x is a number. The extra efficiency *is* significant when working with matrices, though.

For example, suppose A is a 10×10 matrix, and we need to compute

$$p(A) = A^3 + 5A^2 + 3A - 2I.$$

(Here, I is the identity matrix, and $2I$ plays the role of the constant term of the polynomial.) Roughly speaking, multiplying two $n \times n$ matrices A and B is an n^3 operation. Each entry of AB requires n multiplications and $n - 1$ additions, and there are n^2 entries to be found. Thus, computing AB requires on the order of n^3 multiplications and additions of real numbers. If A is 10×10, squaring it takes about 1000 operations; cubing it takes 2000 operations. Therefore, direct evaluation of $p(A)$ involves on the order of 3000 operations. Using Horner evaluation, the computation becomes $((A + 5I)A + 3I)A - 2I$. Only two matrix multiplications occur in this expression — one for each multiplication by A. Therefore, the Horner evaluation is roughly $2/3$ as long as the direct computation in descending form. For large n, the n^3 computation factor becomes gargantuan. Then, a savings of $1/3$ can be very important.

More generally, for a polynomial of degree m, since direct computation of A^k requires $k - 1$ matrix multiplications, evaluating $p(A)$ in descending form takes

$$(m - 1) + (m - 2) + \cdots + 1$$

multiplications. That is on the order of m^2 matrix multiplications for descending form, as compared to order m for Horner's form.

Interestingly, evaluating matrix polynomials via the Horner approach is *not* optimal. There are even more efficient methods known. To illustrate this idea let us look at a very special case: raising a matrix to an integer power.

Suppose we wish to compute A^{11}. In this case, Horner's form and the descending form are the same, assuming the most direct method for computing a power of a matrix is through repeated multiplication. In either, finding A^{11} requires 10 multiplications. But here is a shorter method: square A and square the result to get A^4; now multiply by A obtaining A^5;

square once more to produce A^{10}; and finally multiply again by A to end with A^{11}. That only takes five matrix multiplications, and so saves half the work.

Once you get the idea to use both squaring and multiplications by A, the rationale for the preceding computation is clear enough. Here it is natural to ask for a systematic approach that can be used to find the right combination of squarings and multiplications for any n. And there is a beautiful solution: express the exponent in binary, and then express *that* using Horner evaluation.

In binary, eleven is expressed as

$$1011 = 1 \cdot 2^3 + 0 \cdot 2^2 + 1 \cdot 2 + 1.$$

A polynomial! It can be expressed in Horner form as

$$11 = (((1)2 + 0)2 + 1)2 + 1.$$

Now use this in our original problem:

$$A^{11} = A^{(((1)2+0)2+1)2+1}.$$

Using rules of exponents, this expression can be systematically restructured, working from the outside in. Apply the rule about adding exponents to the final addition of 1. That produces

$$A^{11} = A^{(((1)2+0)2+1)2} \cdot A.$$

Next, looking at the right-most factor of 2 in the exponent, we can use the rule for multiplying exponents. Thus

$$A^{11} = [A^{((1)2+0)2+1}]^2 \cdot A.$$

Carrying this process as far as possible, we get

$$A^{11} = [[A^2]^2 \cdot A]^2 \cdot A.$$

This gives A^{11} through the same sequence described earlier: square, square, multiply by A, square, multiply by A.

The example suggests a general rule. Express the exponent n in binary notation. The left-most digit will be a 1, telling us to begin with A. Then, working from left to right, each digit of 0 tells us to square, while each digit of 1 says to both square and multiply by A. That gives an efficient way to raise A to exponent n. It takes on the order of $\log n$ multiplications because the number of binary digits of n is roughly the base 2 logarithm of n. This is much less than the roughly n multiplications for the direct approach.

This approach to computing powers can be nicely demonstrated on a hand calculator. For example, suppose the goal is to compute 1.23^{11}. Suppose further that your calculator has a memory (accessed with the RECALL key). Enter 1.23 into the calculator memory and then push the following sequence of keys:

$$\boxed{\text{RECALL}}\ \boxed{x^2}\ \boxed{x^2}\ \boxed{\times}\ \boxed{\text{RECALL}}\ \boxed{=}\ \boxed{x^2}\ \boxed{\times}\ \boxed{\text{RECALL}}\ \boxed{=}.$$

That implements the same sequence of squarings and multiplications derived above.

Of course, most calculators used in mathematics classes have a key for the general exponent operation, so the example is uninspiring from a practical standpoint. On the other hand, the far simpler calculators found in most households are typically equipped with a key for squaring (the x^2 key) but no general exponentiation key. Granted most householders seldom need to calculate 1.23^{11} at all, much less to do so efficiently. But if one ever does, it is comforting to know that Horner evaluation is available.

Frequently household calculators also have a square root key, but no general n-th root key, giving rise to questions of the following sort: is there an efficient way to approximate a fifth root (say) using only the square root key? Here is a hint: $1/5$ (or any other proper fraction), can be expressed as the binary equivalent of a decimal expansion. Specifically, $1/5 = .001100110011....$ Here the digits are coefficients of powers of $1/2$ rather than powers of $1/10$.

The method used for an exponent of 11 can be adapted to exponent $1/5$. For now the details will be left to the adventurous reader. More will be said on this subject at the end of the chapter.

So far, we have seen how Horner's form is related to synthetic division, derivatives, efficient polynomial evaluation, and efficient computation of integer and fractional powers. We turn next to an aspect of Horner's form with a different flavor entirely — a beautiful geometric representation for polynomial roots.

1.3   Horner's Form and Lill's Method

I heard about Lill's method at a lecture on mathematics and origami. The speaker, Thomas Hull, described the general method and demonstrated the solution of cubic equations through paper folding. Needless to say, I found the very idea of solving equations by folding paper charming. But even more appealing was the simplicity of the proof of Lill's method. And lurking in the heart of the proof is Horner's form for a polynomial.

Lill's method is actually a bit of a misnomer. It is not a *method* in the sense of telling how to find roots. Rather, it provides a novel way of visualizing them geometrically. Lill, who was an Austrian military engineer, presented his discovery at the Paris Exposition in 1867. Soon after he left the military and became a railway engineer, eventually serving as head of the statistical department for the Austrian Northwest Railway. Little more is known about his contributions, although he apparently discovered *Lill's Law*, of interest in the study of traffic patterns in transportation systems. Today, Lill's method seems to be largely forgotten, despite periodic efforts to publicize it. See the section on historical background at the end of this chapter.

To illustrate Lill's method, let us take the polynomial

$$p(t) = 4t^4 + 6t^3 + 5t^2 + 4t + 1.$$

We use the coefficients to create a polygonal path, as follows. Start at the origin and draw a segment of length 4 along the positive x-axis. Now make a quarter turn counter-clockwise and draw a segment of length 6. Continue this process, successively turning counter-clockwise through right angles and drawing segments, one for each coefficient of p. The result is shown in Fig. 1.1.

We will call this the S-path of polynomial $p(t)$. Denote the k-th side by S_k, where k goes from 0 to 4. Thus S_0 is the horizontal segment starting from the origin and corresponds to the t^4 term of p, S_1 corresponds to the t^3 term, and so on.

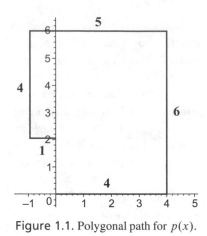

Figure 1.1. Polygonal path for $p(x)$.

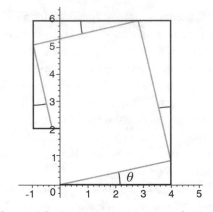

Figure 1.2. T-path defined for angle θ. The T-path is gray; the S-path is black.

Using the S-path as a frame, we now construct a second polygonal path, the T-path. Draw an initial first segment, T_0 from the origin to a point on S_1, and let θ be the angle between S_0 and T_0. For T_1, turn $90°$ counter-clockwise and draw a segment that ends on S_2. Repeat this process drawing T_2 to end on S_3 and T_3 to end on S_4. See Fig. 1.2. Note that all of the marked angles in the figure are equal to θ.

The T-path is variable — it depends on θ. For example, increasing θ slightly in the figure appears to rotate the entire T-path counter-clockwise, although in fact the lengths of the legs T_k will also change. But it seems clear that, with just the right choice of θ, the S and T-paths will end at the same point, as illustrated in Fig. 1.3. For that choice of θ, $t = -\tan\theta$ is a root of $p(t)$. Constructing a root in this way is Lill's method.

For the example at hand, the critical angle is approximately $19.68°$, and the root comes out to about $-.35766$. You can check that $p(-\tan 19.68°)$ is very nearly 0. My calculator gives it as $-.00009062994$.

This geometric condition for a root of p is beautifully simple, and amazingly unexpected. Yet the proof that Lill's method is correct is quite easy. It depends on similar triangles, and the Horner form of the polynomial.

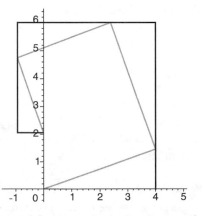

Figure 1.3. A root of p is given by $-\tan\theta$.

Proving Lill's Method

Here is how the proof goes. Consider a generic T-path, as shown in Fig. 1.2. Let $r = -\tan\theta$. We want to see that the S and T paths end at the same point just when r is a root of p.

Together, the S and T paths form a series of similar right triangles, all with a common angle θ. Let us proceed from one triangle to the next, determining the lengths of the legs. As we traverse the S-path starting from the origin, in each triangle we trace first the leg adjacent to the angle θ, and then the leg opposite. It will be convenient to refer to these as the adjacent and opposite legs of the successive triangles.

For the initial triangle, the adjacent leg has length 4, so the length of the opposite leg is $4\tan\theta$. Advancing to the next triangle, the adjacent leg is what remains of side S_1 (outside the first triangle), and so has length $6 - 4\tan\theta = 4r + 6$.

Continuing, the opposite leg of the second triangle las length $(4r + 6)\tan\theta$. That leaves $5 - (4r + 6)\tan\theta = (4r + 6)r + 5$ as the length of the adjacent leg of the third triangle. This pattern continues around the path. Once we know the length of the adjacent leg of one triangle, to find the adjacent leg of the succeeding triangle we multiply by $\tan\theta$ and subtract from the next side of the S-path. Algebraically, that is the same as multiplying by r and adding the next coefficient from p. Thus the successive determination of the adjacent legs of the triangles proceeds by exactly the same computations as the Horner evaluation of $p(r)$.

Carrying this process to its conclusion, we get $(((4r + 6)r + 5)r + 4)r + 1$ for the part of S_4 that extends beyond the opposite leg of the last triangle. But this will be 0 when the condition of Lill's method occurs. Thus, the two paths end on a common point if and only if

$$(((4r + 6)r + 5)r + 4)r + 1 = 0.$$

Since the left side of this equation is the Horner form for $p(r)$, that is the same as saying that $r = -\tan\theta$ is a root of p if and only if the T-path is constructed so that it ends at the same point as the S-path.

Unfortunately, it is not clear in general how to go about finding such a construction. There is a method for general cubics, using geometric operations based on paper-folding. (See the section on additional reading for more information.) This shows that paper-folding operations are more powerful than the classical compass and straightedge methods considered in ancient Greece, which cannot construct roots of cubics in general. But even origami methods cannot be used to construct roots of all polynomials. So, while Lill's method provides an intriguing geometric realization of roots of polynomials, it does not lead to exact geometric methods of solution.

On the other hand, dynamic geometric software can give one the illusion of actually constructing the roots of a polynomial. The idea is to set up the S-path and one sample T-path, and then drag the end of T_0 up and down until the configuration of Lill's method is observed. There are animated graphics illustrating this procedure at the website for this book [87].

Lill's method only locates real roots of a polynomial, because it corresponds to Horner evaluation of $p(r)$ for the real value $r = -\tan\theta$. In fact, for any choice of θ, $p(r)$ gives the (signed) distance between the ends of the S and T paths. Visualize the end of the T path moving along the final leg of the S path, as θ varies. That motion exactly reproduces the projection on the y axis of a point tracing the graph of $y = p(x)$. At x intercepts on this graph the projection passes through the origin, corresponding to configurations in which the S and T paths end at the same point in Lill's construction.

Now suppose p has no real roots, so the graph remains either completely above the x axis or completely below it. In this case, $|p(x)|$ will assume an absolute minimum m. Returning to the earlier visualization, the projection on the y axis of a point tracing the graph of p never gets closer to the origin than m. Likewise, as θ varies in Lill's construction, the end of the T-path never gets closer than m to the end of the S-path. This situation is illustrated in one of the examples presented at the website.

Extensions of Lill's Method

There are several attractive extensions of Lill's method. For one thing there is a nice recursive construction for finding successive roots. Lill's method also provides a new way to think about the quadratic formula. Reversing the direction of the paths leads to another interesting idea. And the method has a more general setting than directly indicated by the presentation earlier.

Successive Roots. For a polynomial p, once we find a root r, we know that $(t - r)$ is a factor of $p(t)$. Thus, we can divide by $(t - r)$, reducing the degree of the polynomial by one, and continue to seek roots. This has a nice realization using Lill's method.

Say we use the method a first time to find a root r. In the process, we generate a corresponding T-path, with the same start and end points as the S-path, but with one fewer legs. Now let the T-path play the role of S-path for a new polynomial, and repeat Lill's method. It turns out that this new polynomial is a constant multiple of $p(t)/(t - r)$, so that reapplying Lill's method is equivalent to reducing the degree as described above. This is illustrated in Fig. 1.4, which shows the secondary T-path as a dashed line. The first root is $-\tan\theta$ and a second root is $-\tan\phi$.

Why does this repeated process work? Look again at the proof of Lill's method, and note the lengths computed for the adjacent legs of the respective triangles. These lengths are, in order, 4, $4r + 6$, $(4r + 6)r + 5$, and $((4r + 6)r + 5)r + 4$, with $r = -\tan\theta$.

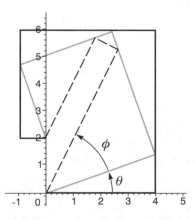

Figure 1.4. Repeated application of Lill's Method.

But these are precisely the coefficients of $p(t)/(t - r)$, as we found in the discussion of synthetic division. The T-path is made up of the hypotenuses of the right triangles, and each hypotenuse is $\sec \theta$ times the corresponding adjacent leg. This shows that the T-path in our original diagram is precisely the S-path for the polynomial $\sec \theta \cdot p(t)/(t - r)$.

The Quadratic Formula.   Let us use Lill's method to derive the quadratic formula. Given the quadratic $at^2 + bt + c$, with positive cofficients, we create the diagram in Fig. 1.5, corresponding to one root $r = -\tan \theta$. The S and T-paths both start at the origin O and end at $Q = (a - c, b)$. We also mark point R where the T-path has its right angle. The coordinates of R are $(a, a \tan \theta) = (a, -ar)$.

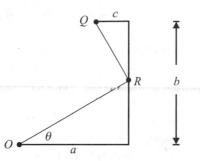

Figure 1.5. Lill's method for a quadratic polynomial.

A circle having OQ as a diameter must also pass through R, because there is a unique circle through any three noncollinear points, and because a right angle at any point of a circle subtends a diameter. The center of the circle will be at the midpoint of OQ, $(\frac{a-c}{2}, \frac{b}{2})$; its radius is half the length of OQ. Thus, the equation of the circle is

$$\left(x - \frac{a - c}{2}\right)^2 + \left(y - \frac{b}{2}\right)^2 = \frac{b^2 + (a - c)^2}{4}.$$

But $(a, -ar)$ lies on this circle. Substituting,

$$\left(\frac{a + c}{2}\right)^2 + \left(ra + \frac{b}{2}\right)^2 = \frac{b^2 + (a - c)^2}{4}.$$

A little algebra gives

$$\left(ra + \frac{b}{2}\right)^2 = \frac{b^2 - 4ac}{4},$$

and solving for r produces the quadratic formula,

$$r = \frac{-b \pm \sqrt{b^2 - 4ac}}{2a}.$$

Reversed Coefficients.   Another nice extension of Lill's method is to reverse direction in drawing the two paths. When the angle is properly selected to determine a root of p, the two paths start at the same point and end at the same point. So what happens if we interchange the roles of starting and ending points? We find a Lill's diagram for a root of a new polynomial, with coefficients given in reverse order. The turns are all clockwise instead of counter-clockwise, but that doesn't matter — the verification of Lill's method is still valid. However, when we reverse direction, we shift from what was originally the angle θ to the complimentary angle. Consequently, the root for the new polynomial is the reciprocal of the root for the original polynomial. In this way, the Lill approach provides a nice visual demonstration for a fairly well-known fact: reversing the order of the coefficients of a polynomial has the effect of inverting the nonzero roots.

Reversing directions in the Lill's diagram also ties back to the Horner evaluation of polynomials. As an illustration of this idea, we will look again at the particular polynomial

$$p(t) = 4t^4 + 6t^3 + 5t^2 + 4t + 1.$$

Using the Horner form to compute $p(r)$ for any r involves repeatedly adding coefficients and multiplying by r. In fact, starting with 0, we add the first coefficient (4) then multiply by r; add 6 and multiply by r; and continue this process until we have added the final 1.

Whatever the result, we can reverse the process by undoing all of these steps in reverse order: subtract 1 and divide by r; subtract 4 and divide by r again. Repeat this process until we have subtracted the initial 4, and we will be back at the starting point, namely 0.

Now suppose that r is a root of p. After the forward Horner evaluation we will reach a result of 0. In this case, the reverse process is again a Horner evaluation. It starts with 0 and successively adds constants (the negatives of the coefficients) and multiplies by a fixed value $(1/r)$. That is the Horner evaluation for the polynomial

$$q(t) = -t^4 - 4t^3 - 5t^2 - 6t - 4$$

at $t = 1/r$. And since we know it results in 0, $1/r$ is a root of $q(t)$. That also shows that $1/r$ is a root of $-q(t)$, which is the reverse polynomial for p.

As mentioned, the fact that the (nonzero) roots of the reverse polynomial are the reciprocals of the (nonzero) roots of the original polynomial is fairly well known. It is easy to verify directly by simply evaluating the reverse polynomial (in descending form) at $1/r$, assuming r is a root of the original polynomial. The combination of Lill's method and Horner evaluation gives a nice alternate way to think about this result. We will see reverse polynomials again in Chapter 3.

General Formulation.   Our final extension of Lill's method, and one that really cannot be overlooked, is prompted by the special properties of the model polynomial and its diagram in Fig. 1.3. For example, all the coefficients are positive, and the S-path never crosses itself. Also, the model example says to draw T_0 from the origin to a point on S_1. This imposes a restriction on the values we can consider for θ, and hence for a possible root r. Do we really need such special properties? Or can we see that Lill's method works for any polynomial?

It is convenient here to use vector methods. We will begin by defining the sides of the S-path. To arrange for the appropriate right angle turns, let us define R to be the linear transformation that rotates the plane 90° counter-clockwise about the origin. It is possible to express R using a 2×2 matrix, but that is not explicitly needed. It suffices to observe that application of R amounts to multiplying by a fixed matrix, and so the usual rules of algebra apply.

Given a polynomial

$$p(t) = a_0 t^n + a_1 t^{n-1} + \cdots + a_n,$$

we can now specify the sides comprising the S-path. Define

$$S_n = a_n R^n \mathbf{e} \tag{3}$$

where $\mathbf{e}$ is the unit vector in the direction of the positive x axis. Actually, we might take any fixed vector for $\mathbf{e}$, the choice of which defines the direction of the initial side of the S-path. Visualizing the plane as a map, our choice of $\mathbf{e}$ makes the starting direction due east. The successive directions, $R\mathbf{e}$, $R^2\mathbf{e}$, $R^3\mathbf{e}$, will then be north, west, south, and so on.

Now we partition each vector S_k into two parts, A_k and B_k, separated by the point where the T-path will touch S_k (see Fig. 1.6). Thus,

$$S_k = A_k + B_k. \tag{4}$$

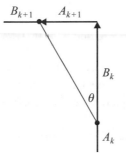

Figure 1.6. Successive legs of the S-path. $S_k = A_k + B_k$; $S_{k+1} = A_{k+1} + B_{k+1}$.

Consistent with the earlier description of Lill's method, we take

$$A_{k+1} = (\tan \theta) R B_k. \tag{5}$$

In contrast with the original description of Lill's method, in which the T-path was defined in terms of θ and the dependence of A_{k+1} on B_k arose as a consequence, here we take (5) as the defining property. Then, once we agree that $A_0 = 0$, all the remaining A's and B's are determined inductively by (4) and (5). In turn, the T-path is created by defining the sides as vectors $T_k = B_k + A_{k+1}$.

As before, the goal is to choose θ so that $B_n = 0$. We will see that this occurs if and only if $r = -\tan \theta$ is a root of the polynomial p. Accordingly, let us rewrite (5) as

$$A_{k+1} = -rRB_k.$$

Combined with (4), this gives us

$$S_{k+1} - B_{k+1} = -rRB_k,$$

and after rearrangement

$$B_{k+1} = rRB_k + S_{k+1}. \tag{6}$$

This is the equation we will use to determine all of the B's.

From the starting assumption $A_0 = 0$ we immediately obtain $B_0 = S_0 = a_0\mathbf{e}$. Now apply (6) with $k = 0$. That gives us

$$
\begin{aligned}
B_1 &= rRB_0 + S_1 \\
&= rRa_0\mathbf{e} + a_1 R\mathbf{e} \\
&= R(a_0 r + a_1)\mathbf{e}.
\end{aligned}
$$

Here, the algebraic manipulations are justified by properties of matrix, vector, and scalar operations. It is worth observing that they remain valid for any 2×2 matrix R.

Repeating the process with $k = 2$, we find

$$
\begin{aligned}
B_2 &= rRB_1 + S_2 \\
&= rR(R(a_0 r + a_1)\mathbf{e}) + a_2 R^2\mathbf{e} \\
&= rR^2(a_0 r + a_1)\mathbf{e} + R^2 a_2\mathbf{e} \\
&= R^2((a_0 r + a_1)r + a_2)\mathbf{e}.
\end{aligned}
$$

The pattern is evident. Each time we apply (6), we obtain an expression of the same form: a power of R, one of the intermediate results in the Horner evaluation of $p(r)$, and the vector $\mathbf{e}$. We can see that the pattern persists: for $k = 1, 2$ we have

$$
B_k = R^k q(r)\mathbf{e},
$$

where $q(r)$ is a polynomial in r. When we apply (6) we get

$$
\begin{aligned}
B_{k+1} &= rRB_k + S_k \\
&= rR(R^k q(r)\mathbf{e}) + a_{k+1} R^{k+1}\mathbf{e} \\
&= R^{k+1}(rq(r) + a_{k+1})\mathbf{e}.
\end{aligned}
$$

This shows that each succeeding B_k will be in the same form. Moreover, at each step we obtain the new $q(r)$ by multiplying by r and adding the next coefficient of $p(t)$, exactly the algorithm for Horner evaluation. Therefore, the pattern we noticed at the start continues and at the end, we will have

$$
B_n = R^n p(r)\mathbf{e}.
$$

Thus B_n is 0 if and only if $p(r) = 0$. This shows that the Lill method correctly characterizes a root of $p(t)$ in the more general setting.

This development leads to two additional points about Lill's method. First, there is no reason why R must be a rotation of $90°$. Any angle ϕ between 0 and $180°$ works just as well. We have to redefine $r = -\sin(\theta)/\sin(\phi - \theta)$, which reduces to the original definition of r when $\phi = 90°$ (see Fig. 1.7). As the figure shows, we will still obtain the equation

$$
A_{k+1} = -rRB_k,
$$

although in this instance it follows from the law of sines.

With these modifications, the derivation works exactly as before. This shows that Lill's method works even for non-perpendicular paths. If the T path starts out at an angle θ to the

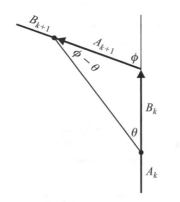

Figure 1.7. A and B vectors for a non-perpendicular path.

S path, and if the two paths end at the same point, then $r = -\sin(\theta)/\sin(\phi - \theta)$ is a root of the polynomial. As an illustration, Fig. 1.8 shows two Lill diagrams for the polynomial

$$p(t) = 4t^4 + 6t^3 + 5t^2 + 4t + 1.$$

In the diagram on the left, each path has perpendicular sides. In the diagram on the right, R is a rotation through $80°$. For each diagram, θ has been chosen to produce the same root r.

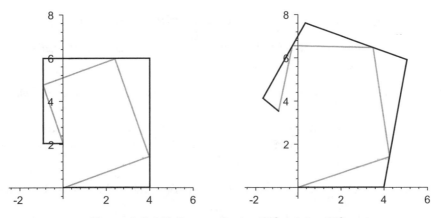

Figure 1.8. Lill diagrams for $\phi = 90°$ and $\phi = 80°$.

The second additional point about Lill's method concerns T paths that stray away from the S path. Notice that in drawing the diagram for the general version of Lill's method, the vertices of the T-path need not all fall within the segments making up the S-path. The algebraic proof above works perfectly with no restriction on the value of θ. But it means that we have to take a slightly modified view of how the T-path is to be constructed. The S-path is defined by (3). But now consider the entire line L_k which contains the segment S_k. Then each segment T_k can end anywhere on L_{k+1}, not just within the endpoints of the original segment S_{k+1}. In fact, the general description of the T-path can be given as follows: Start at the origin, and draw a segment T to any point of the line containing S_1. From there, traveling perpendicular to T, find the point of intersection with the line containing S_2. Repeat this process until a final intersection is reached with the line containing S_n. That gives a T-path that works properly for Lill's method, meaning that when the T-path and the S-path end at the same point, $-\tan\theta$ is a root of $p(t)$. But the T-path need not remain within the region bounded by the S-path.

Both of these additional points are demonstrated with animated graphics at the webpage for this book [87].

This brings to a close the discussion of Lill's method, and of Horner's form, although there remains one final section with historical references and suggestions for additional reading. As we have seen, Horner's form is connected to quite a few other topics, including efficient computation, synthetic division, and differentiation. But of all of the topics connected to Horner's form, by far my favorite is Lill's method. It is simple, visual, and unexpected, and it has many appealing twists and turns (so to speak). So inspirational an idea deserves to be celebrated. As a possible step in that direction, I offer the following mnemonic verse:

> Lill, backed by Horner
> Works in a corner
> Angled a fraction of π.
> Sublime rule of thumb:
> When ends become one
> A root lies revealed to the eye.

1.4   History, References, and Additional Reading

Expressing a polynomial in Horner's form has also been referred to as Horner's rule. Regarding its history, Knuth [102, p. 467] has this to say:

> W. G. Horner gave this rule early in the nineteenth century [*Philosophical Transactions*, Royal Society of London, 109 (1819), 308–335] in connection with a procedure for calculating polynomial roots. The fame of the latter method ... accounts for the fact that Horner's name has been attached to [the rule], but actually Isaac Newton had made use of the same idea 150 years earlier.

Knuth goes on to give the specific reference from 1669 where Newton used the Horner form.

The root finding procedure to which Knuth refers was known well before Horner's day. According to Boyer [25, p. 204], the method was well known in 13th century China, appearing in the work of at least three mathematicians of that period. Boyer also mentions that the same method was known to Arabic mathematicians in the 15th century. Of course, these earliest applications of the method were not expressed in our modern algebraic notation. But in specific numerical examples the underlying algorithm can be understood, and this was essentially the same as that used by Horner. It is a reasonable inference, as well, that the practitioners of this method used Horner evaluation as part of the algorithm. Accordingly, what long established custom attributes to Horner actually was known centuries earlier.

Efficient Calculation.   Efficiently computing A^n by expressing n in binary form, and hence as a polynomial in Horner's form, is discussed by Golub and Van Loan [58] under the heading *binary powering*. There the use of Horner evaluation for matrix polynomials is analyzed, and it is shown that Horner evaluation is not optimal. That is a good starting point for further reading on this topic, and provides additional references.

Approximating nth roots. Horner's form and efficient computation of powers is also discussed in my article with Page [90]. This article discusses both nth powers and nth roots. Recall that this idea was touched on earlier in this chapter (see page 13). In fulfillment of a promise made there, here is an illustrative example approximating fifth roots.

Observing the equivalence of *fifth root* and *one-fifth power*, we begin by expressing $1/5$ in binary, that is, as a sum of powers of $1/2$:

$$\frac{1}{5} = \frac{1}{2^3} + \frac{1}{2^4} + \frac{1}{2^7} + \frac{1}{2^8} + \frac{1}{2^{11}} + \frac{1}{2^{12}} + \cdots$$
$$= .001100110011\cdots .$$

If we take

$$\frac{1}{5} \approx .0011 = 0\left(\frac{1}{2}\right) + 0\left(\frac{1}{2}\right)^2 + 1\left(\frac{1}{2}\right)^3 + 1\left(\frac{1}{2}\right)^4$$

we have an approximation for $1/5$ as a polynomial in powers of $1/2$.

Now our goal is to approximate a fifth root, which is a number raised to an exponent of $1/5$. Replace that exponent with its polynomial approximation, expressed using the Horner form. In analogy with the earlier discussion with exponent 11 (page 13) we are led to a sequence of computations involving square roots and multiplications. For example, to approximate the fifth root of 1.23, we would compute

$$1.23^{(((.5+1).5).5).5)))} = ((((1.23^{.5})1.23)^{.5})^{.5})^{.5}.$$

Lill's method. As I mentioned before, I first heard about Lill's method in Hull's talk on the mathematics of paper folding. For additional information about this subject, I recommend Hull's article [70] on angle trisection via paper folding, as well as his book [69] of activities for exploring mathematical ideas using origami. Hull also maintains a website [71] devoted to mathematics and origami, rich with opportunities for further exploration.

Biographical information on Lill was kindly sent to me by Douglas Rogers, who found (and translated) a brief account in an Austrian biographical dictionary [111].

Lill's method has been mentioned in several works, dating almost immediately from his original publication. Among the first was a pamphlet published in 1879 by William Bixby, a Lieutenant of Engineers in the U.S. Army. Part of this pamphlet was reprinted in the *American Mathematical Monthly* in 1922, accompanied by Bixby's recollection

> I read of [Lill's method] about 1878, and published it in 1879 by a privately printed pamphlet. At that date I had not seen Lill's 1867 printed article.... Luigi Cremona had also described Lill's method and made it public to English readers in 1888. My pamphlet failed to attract much attention. [18]

I found Bixby's original pamphlet in the Artemis Martin collection at American University. It is now available electronically at the website for this book [87].

Lill's method was also mentioned by eminent mathematicians L. E. Dickson [40] in 1904 and Felix Klein [98] in 1911. Other accounts date to 1925 [121], 1945 [53], 1952 [63], and 1962 [134].

Bradford [26] has an extensive website devoted to Lill's method. There you will find ideas very closely related to many of the extensions of Lill's method presented above. The website also has other ideas not considered here, including constructive methods for solving cubics and extensions to complex solutions of quadratics.

2
Polynomial Potpourri

The mathematical explorer will find a wide variety of interesting rambles into the polynomial back country, ranging from short day hikes to ambitious treks of days or even weeks. In the preceding chapter we lingered in one locale, becoming thoroughly oriented to the landscape of Horner's form. For this chapter, we will cover more ground, embarking on short excursions in widely scattered directions. This will provide an interlude before the focussed effort of the two succeeding chapters, in which we will mount a determined assault on the solution of polynomial equations. So take your time, set a relaxed pace, and keep an eye out for the unusual and unexpected. I hope you will find many memorable vistas worthy of including in your mathematical scrapbook.

2.1 A Perplexing Puzzle

The owl and the pussycat were playing at riddles one day. The owl said, "I am thinking of a polynomial, $p(x)$. All of its coefficients are whole numbers. Can you guess what it is?"

"Not without a hint," answered the pussycat. "Pray tell me, what is $p(1)$?"

The owl readily complied. Noting the answer down, the pussycat said, "Good. Now be so kind as to tell me $p(p(1))$."

The owl provided that as well. Then, after a little scratch work, the pussycat stated the exact definition of $p(x)$ in descending form. How can this be explained?

This puzzle has appeared in various guises in print and on line. One source is [95]. A solution will be given at the end of the chapter.

2.2 The Quadratic Formula, Insideout

Anyone who has studied algebra knows about the quadratic formula. And it is common knowledge that a monic quadratic with roots r and s can be written $(x - r)(x - s) = x^2 - (r + s)x + rs$. But when these two simple ideas are combined, something unexpected pops up.

To see what that something is, apply the quadratic formula to $x^2 - (r + s)x + rs$, even though we already know the roots. The result is

$$\text{roots} = \frac{r + s \pm \sqrt{(r + s)^2 - 4rs}}{2}$$

$$= \frac{r + s \pm \sqrt{r^2 + 2rs + s^2 - 4rs}}{2}$$

$$= \frac{r + s \pm \sqrt{(r - s)^2}}{2}.$$

It is tempting to simplify that last square root to $r - s$, but that is not quite right. The rule is $\sqrt{a^2} = |a|$, because a can be negative, but $\sqrt{a^2}$ cannot. Therefore we get

$$\text{roots} = \frac{r + s \pm |r - s|}{2}. \tag{1}$$

How can this be reconciled with the fact that the roots are r and s? The answer is tied up with that $\pm$ sign. When it is taken as $+$, we must obtain a larger answer than when it is taken as $-$. Thus, (1) becomes

$$\text{larger root} = \frac{r + s + |r - s|}{2}$$

$$\text{smaller root} = \frac{r + s - |r - s|}{2}.$$

In this way, we have derived well-known formulas for the maximum and minimum of two variables from the quadratic formula.

2.3   Solving Cubics with Curly Roots

Just as the quadratic formula provides solutions to quadratic equations, so there is a cubic formula for cubic equations. For an equation of the form $x^3 + ax + b = 0$, one root is given by

$$\sqrt[3]{\frac{-b}{2} + \sqrt{\frac{b^2}{4} + \frac{a^3}{27}}} + \sqrt[3]{\frac{-b}{2} - \sqrt{\frac{b^2}{4} + \frac{a^3}{27}}}. \tag{2}$$

As we will see in Chapter 3, every cubic equation can be reduced to the form above, so this cubic formula can be used to find roots of any cubic. But look how complicated it is! And for some equations $b^2/4 + a^3/27 < 0$, so that using the formula requires finding a cube root of a complex (non-real) number. Surely there must be a better way.

If you insist on solving your equations in terms of traditional radicals, that is, square roots, cube roots, and so forth, then (2) is about the best you can do. But why should we limit ourselves to traditional radicals? We will see that solving cubics can be a breeze, using a new function I call the *curly root*, denoted $\{\overline{x}$. (The name is inspired by the notation: a radical-like symbol built using a curly brace.) The method is based on work of Nogrady [123] in 1937 and Pettit [130] in 1947, but it may have been suggested by others even earlier.

In general, a radical is an inverse function, as the cube root function is the inverse of $f(x) = x^3$. This can be expressed as

$$y = \sqrt[3]{x} \text{ if and only if } x = y^3.$$

Analogously, we define the curly root function as the inverse of $f(x) = x^3/(1-x)$, so that

$$y = \{\overline{x} \text{ if and only if } x = \frac{y^3}{1-y}.$$

But we have to take care to avoid an ambiguity. The function $x^3/(1-x)$ is not one-to-one on its domain, with up to three values of x producing a single y. To address this problem, we restrict x to the domain $A = \{x| -3 \le x < 1\} \cup \{x| 1 < x < 3/2\}$. Then $x^3/(1-x)$ is a one-to-one function whose range is the entire real line. Likewise, if we restrict $\{\overline{x}$ to lie in A, then it is uniquely defined for all real x. A graph of $\{\overline{x}$ is shown in Fig. 2.1.

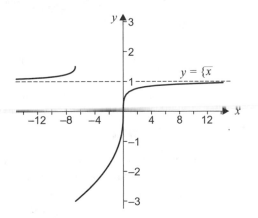

Figure 2.1. Graph of the curly root function.

Though $\{\overline{x}$ does not have the properties of radicals that are so useful in algebraic manipulation, it is a legitimate function and every bit as respectable as $\sqrt[3]{x}$. Furthermore, it does satisfy one important identity:

$$\{\overline{x}^3 = x - x\{\overline{x} \tag{3}$$

In fact, with $y = \{\overline{x}$, this identity says $y^3 = x(1-y)$, which evidently follows from the definition of $\{\overline{x}$.

But the main virtue of $\{\overline{x}$ is that it can be used to find roots of any cubic equation. As we will see in Chapter 3, any cubic polynomial can be reduced to one of the form $p(x) = x^3 + ax + b$. Then one root will be given by

$$r = -\frac{b}{a}\left\{\overline{\frac{a^3}{b^2}}\right..$$

To verify this, substitute in p to find

$$r^3 + ar + b = -\frac{b^3}{a^3}\left(\left\{\overline{\frac{a^3}{b^2}}\right.\right)^3 - a\frac{b}{a}\left\{\overline{\frac{a^3}{b^2}}\right. + b.$$

Using (3), this becomes

$$r^3 + ar + b = -\frac{b^3}{a^3}\left(\frac{a^3}{b^2} - \frac{a^3}{b^2}\left\{\overline{\frac{a^3}{b^2}}\right.\right) - b\left\{\overline{\frac{a^3}{b^2}}\right. + b.$$

Simplifying now leads to

$$r^3 + ar + b = 0,$$

showing that r is a root of p.

To obtain the remaining roots, divide $p(x)$ by $(x - r)$ and then apply the quadratic formula to the result. This gives

$$\frac{b}{2a} \sqrt[\{]{\frac{a^3}{b^2}} \left(1 \pm \sqrt{\frac{\{a^3/b^2\} + 3}{\{a^3/b^2\} - 1}} \right)$$

for the other two roots of p.

You may wonder how to evaluate curly roots. Square roots are found using the square root key on a calculator, and anyone fortunate enough to have a calculator with a curly root key can deal with curly roots similarly. Otherwise, the alternate methods available for evaluating square roots, such as Newton's method, power series, and graphical analysis, can also be used to compute curly roots. The details are left for the interested reader.

2.4   Chaos in Newton's Method

Speaking of Newton's method, did you know that it can lead to chaos[1]? This may seem surprising, given how well behaved the method is most of the time, especially when applied to polynomials. But even in that context chaos does indeed arise. The following discussion shows how, following a few of the key ideas from a terrific paper by Saari and Urenko [139].

Here is a quick review of Newton's method. The goal is to find a root (or an x intercept) of a function $y = f(x)$ through a process of successive approximation. Graphically, the method can be described as follows: start with an x near the intercept, locate the corresponding point $(x, f(x))$ on the graph, slide along the tangent line to reach a new x even closer to the intercept, repeat. This idea is dramatized with animated graphics at the website for this book [87]. A screen image is shown in Fig. 2.2.

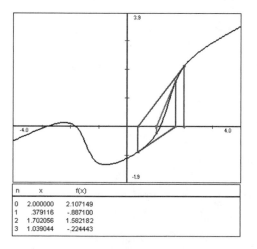

Figure 2.2. Sliding down the tangent line.

---

[1]Here, *chaos* has its mathematical meaning, as in [38, 39, 61].

It is an easy exercise to derive the analytic version of Newton's method,

$$x_{n+1} = x_n - \frac{f(x_n)}{f'(x_n)},$$

which specifies how to go from one value x_n to the next. From an x_0 sufficiently close
to a root of $f(x)$, it can be shown that the sequence $\{x_n\}$ converges rapidly to the root.
However, chaos arises as soon as we consider what happens if x_0 is *not* sufficiently near
the root.

Suppose that $f(x)$ is a polynomial with degree m, m distinct real roots $r_1, r_2, \ldots, r_m$,
and a leading coefficient of 1. An example is graphed in Fig. 2.3. For this f we define
the Newton's function $N(x) = x - f(x)/f'(x)$. For any starting value x_0, Newton's
method produces the sequence $x_0, N(x_0), N(N(x_0)), \ldots$. Thus, the sequence has nth term
$x_n = N^{(n)}(x_0)$, meaning that x_n is found by applying N n times.

Figure 2.3. A typical polynomial for Newton's method.

Newton's method is an example of function iteration, which is one place chaos can arise.
We can consider three sets for any function: the initial points for which the sequence of
iterates converges, the points for which it diverges to $\pm\infty$ (which is empty for $N(x)$); and
the complementary set of points for which the sequence does neither. It is in connection
with this last set that the interesting phenomena occur.

To help analyze the behavior of $N(x)$ under iteration, we make a number of observations. First, from our assumptions, the derivative $f'(x)$ will have $m - 1$ distinct real roots
$s_1, s_2, \ldots, s_{m-1}$, interlaced between the roots of f. Second, N is a rational function with a
pole or vertical asymptote at each s_j. Third, $N(x) = x$ if and only if $f(x) = 0$, so that the
roots of f are the fixed points of N. These features are visible in the graph of N in Fig. 2.4.

The graph provides additional insight about N. First, the fixed points of N occur where
the curve $y = N(x)$ intersects the line $y = x$. Second, the graph of N is partitioned
by vertical asymptotes, one for each root of f', so the domain of N is the union of open
intervals. Near their endpoints $|N(x)|$ approaches infinity; within each there is just one root
of $f(x)$. Of these intervals, $m - 2$ have finite length, and within each of these $N(x)$ ranges
from $-\infty$ to ∞. Let us call the intervals making up the domain of N the *fundamental
intervals* for N, and the ones of finite length *full-range* intervals. If the r_k and s_k are

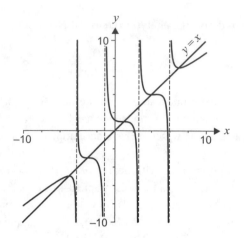

Figure 2.4. The graph of N with the line $y = x$.

ordered least to greatest, then the fundamental intervals will be $I_1 = (-\infty, s_1)$, $I_2 = (s_1, s_2), \dots, I_m = (s_{m-1}, \infty)$, and $r_k \in I_k$ for each k.

For an initial value x_0, the sequence of iterates $x_n = N^{(n)}(x_0)$ has an appealing visualization using the graph of N. Begin with x_0 on the x-axis. Draw a vertical line to the curve $y = N(x)$, ending at $(x_0, N(x_0)) = (x_0, x_1)$. Next, draw a horizontal line to $y = x$, reaching the point (x_1, x_1). Now repeat the process: draw a vertical line to the curve $y = N(x)$ and then a horizontal line to $y = x$, ending at (x_2, x_2). Continuing we can observe the sequence of iterates as we trace a rectangular path. Adding this path to our graph produces a *cobweb diagram* , so called because the path often takes the form of a rectangular spiral, resembling a spider web. In Fig. 2.5 the first segments in a cobweb diagram are shown for two different values of x_0.

On the left the sequence converges to a limit, which is what we expect when using Newton's method. The cobweb path takes the shape of an inward spiral, which is typical when the graph of N is decreasing as it crosses $y = x$. Where it is increasing, a convergent sequence will approach its limit monotonically, forming a cobweb path that looks like a staircase. In either case, when the Newton sequence converges, its terms eventually remain in a single fundamental interval I_k, and the limit is the root of f in that interval. The diagram on the right shows how, if x_0 is not sufficiently close to the root, Newton's method can produce terms that cross from one fundamental interval to another.

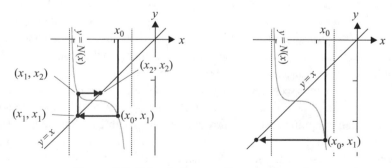

Figure 2.5. Cobweb diagrams for two values of x_0.

Looking at cobweb diagrams at various points along the graph of N shows that in each fundamental interval I_k there is a subinterval of points x_0 from which the Newton sequence converges to r_k. If a Newton sequence ever enters the subinterval it is captured, converging to f's unique root in I_k. But not all sequences meet this fate. For some x_0 the Newton sequence never converges. The range of possibilities for the nonconvergent sequences is surprisingly rich.

In general, Newton sequences for a polynomial f never go to infinity or minus infinity. Given our assumptions about f, the convergence subintervals within I_1 and I_m extend to $-\infty$ and ∞, respectively. Thus, for any x_0, if some $|x_n|$ is sufficiently large, the sequence is captured by a convergence subinterval and tends to a finite limit. In particular, each x_0 generates a bounded sequence.

How does such a sequence behave? If it does not converge, it might cycle through a finite set of values. Suppose that $N^{(k)} x_0 = x_0$ for some k. Then, after k steps, the sequence $\{x_n\}$ returns to its starting point, and so repeats the same cycle again. In this case x_0 is called a periodic point of period k. Analyzing the graph of N shows that there are periodic points of all periods, with the number of points of period k increasing exponentially with k. A large abundance of periodic points is a characteristic of chaos.

The animated demonstration of Newton's method at the website for this book [87] illustrates periodic behavior. In one example, x_0 is near a periodic point of period 6. It generates a sequence that jumps around among the fundamental intervals, almost forms a closed cycle, but eventually settles down and converges to one root. Changing the value of x_0 in the 12th decimal place produces nearly identical behavior, but in the end the sequence converges to a different root. See Fig. 2.6. This is an example of *sensitive dependence on initial conditions*, another characteristic of chaos. The idea is that a microscopic change at the start of the process can lead to a dramatic difference in the final outcome.

There remains another possible behavior for a nonconvergent Newton sequence. It can jump around erratically from one fundamental interval to another without ever settling down in one interval. Indeed, if we take any sequence whatsoever of full-range funda-

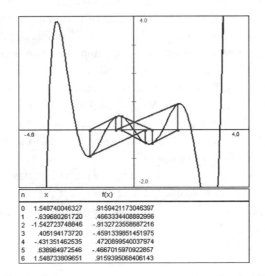

n	x	f(x)
0	1.548740046327	.9159421173046397
1	-.639680261720	.4663334408892996
2	-1.542723748846	-.9132723558687216
3	.405194173720	-.4591339851451975
4	-.431351462535	.4720899540037974
5	.638964972546	-.4667015970922857
6	1.548733809651	.9159395068406143

Figure 2.6. Newton sequence near a point of period 7.

mental intervals, for example $I_4, I_2, I_2, I_4, I_3, \ldots$, there is some x_0 for which the Newton sequence falls into the fundamental intervals in precisely the specified order. The possibility of such erratic behavior is another characteristic of chaos, and doubtless inspires the very name.

All of these assertions can be understood by visualizing the graphs of iterates of N. That is, imagine graphing $y = N(N(x))$, $y = N(N(N(x)))$, and so on. Consider a finite fundamental interval, I, in Fig. 2.4. As x varies over I, $N(x)$ ranges through the entire real line. Therefore, applying $N(N(x))$ to the points in I produces a graph like that of $N(x)$ over all reals. We can envision the new graph as the full graph of N compressed horizontally to fit between the endpoints of I. It will also be reflected across a vertical line, because the graph of N restricted to I is decreasing, going from infinity to negative infinity as x traverses I from left to right. But the key point is that the graph of $N^{(2)}$ will have the same number of vertical asymptotes, fundamental intervals, and corresponding convergent subintervals within I as the full graph of N has over $\mathbb{R}$. A similar statement can be made for each fundamental interval. Thus, the graph of $N^{(2)}$ will have m^2 fundamental intervals, and within most of them (all of the sections that come from finite fundamental intervals of N) the range will be the entire real line. There are $m - 2$ full-range intervals for N, and at least $(m - 2)^2$ for $N^{(2)}$.

Now we can easily visualize the location of periodic points with period 2. They satisfy $N^{(2)}(x) = x$ and so correspond to intersections of the graph of $N^{(2)}$ and the line $y = x$. But there must be an intersection within each of the full-range intervals for $N^{(2)}$. Therefore, there will be at least one point of period 2 in each full-range interval of $N^{(2)}$, and hence at least $(m - 2)^2$ period 2 points overall.

Apply these same ideas to the higher iterates $N^{(n)}$. Within each full-range interval I_k, the graph of $N^{(2)}$ has a compressed version of the full graph of N, partitioned into its own small fundamental intervals. Similarly, within each of the full-range fundamental intervals of $N^{(2)}$, the graph of $N^{(3)}$ will appear to be an even more tightly compressed copy of the full graph of N. Continuing, the graph of $N^{(k)}$ will have m^k fundamental intervals, of which $(m-2)^k$ will be full range, so $y = x$ intersects the graph of $N^{(k)}$ in at least $(m-2)^k$ points. Thus there will be at least $(m - 2)^k$ points of period k. As an example, the f in Fig. 2.3 has degree $m = 5$. Therefore, there are at least 3^8 points of period 8.

We can also see how Newton's sequences can jump from one fundamental interval to another. For example, let us consider J, the second fundamental interval for $N^{(2)}$ within N's fundamental interval I_3. Suppose that x_0 is an element of J, and so in I_3. We can see that $x_1 = N(x_0)$ will be in I_2. Applying $N^{(2)}$ to the elements of I_3 produces a compressed version of the graph of N. The interval J consists precisely of the points that fall into I_2 after one application of N. This shows that any x_0 in J leads to a Newton sequence that starts in I_3 and then immediately hops to I_2.

Suppose we want a Newton sequence that starts in I_3, and then travels to I_2, I_4, and back to I_2. Inside I_3 find the second fundamental interval $I_{3,2}$ for $N^{(2)}$. Within that interval find the fourth fundamental interval $I_{3,2,4}$ for $N^{(3)}$. Finally, within $I_{3,2,4}$ consider the second fundamental interval $I_{3,2,4,2}$ for $N^{(4)}$. Every x_0 in the last interval generates a Newton sequence that visits the fundamental intervals I_k in the order specified. Applying the same construction with n prespecified intervals I_k produces a nested sequence of n intervals. Any x_0 in their intersection produces a Newton sequence that visits the fundamental intervals in the specified order. It is possible to extend this idea to an infinite sequence of I_k's using properties of intersections of nested intervals.

The goal of this discussion has been to provide an account of the range of possibilities that arise when Newton's method is applied to a polynomial. Considering all possible initial points x_0, several characteristic properties of chaotic systems occur, including an abundance of periodic points, sensitive dependence on initial conditions, and sequences with erratic behavior. I have attempted to give an overview of these ideas, without going into technical details or rigorous proofs. The paper by Saari and Urenko [139] gives a mathematically complete treatment and includes additional results about chaotic behavior and the sets where it is observed.

2.5   Polynomial Interpolation

Everybody knows that two points determine a line. This idea generalizes to more points and higher degree polynomials: three points determine a quadratic, four a cubic, five a quartic, and so on. The polynomial is said to *interpolate* the given points. In general k points with distinct x coordinates determine a unique interpolating polynomial of degree $k - 1$.

There is a simple general formula for an interpolating polynomial, referred to as the *Lagrange interpolating polynomial*, although reportedly Lagrange was not the first to recognize it (see [4]).

An example will show how the Lagrange interpolating polynomial is determined. Suppose we wish to find a polynomial whose graph passes through $(-1, 4)$, $(2, 5)$, $(3, -2)$, and $(4, 1)$. Look at the four x values. We can find a polynomial that has roots at three of them and passes through the specified point for the fourth. Let $q_1(x) = (x - 2)(x - 3)(x - 4)$. Then $q_1(x)/q_1(-1)$ has roots at $x = 2, 3$, and 4, and takes the value 1 at $x = -1$. Therefore, $p_1(x) = 4q_1(x)/q_1(-1)$ passes through the point $(-1, 4)$, and is 0 at the other x values. In the same way, we can find polynomials $p_j(x)$ for $j = 2, 3, 4$, so that p_j passes through the jth point, and is 0 at the x coordinates of the remaining points. Then $p_1 + p_2 + p_3 + p_4$ is a polynomial that passes through the four points. Thus we have the following formula for the interpolating polynomial:

$$p(x) = 4 \frac{(x - 2)(x - 3)(x - 4)}{(-1 - 2)(-1 - 3)(-1 - 4)} + 5 \frac{(x + 1)(x - 3)(x - 4)}{(2 + 1)(2 - 3)(2 - 4)}$$
$$- 2 \frac{(x + 1)(x - 2)(x - 4)}{(3 + 1)(3 - 2)(3 - 4)} + \frac{(x + 1)(x - 2)(x - 3)}{(4 + 1)(4 - 2)(4 - 3)}.$$

The example shows the general pattern. If the points we are to interpolate are (x_k, y_k), $k = 1, 2, \ldots, n$, then we define polynomials $q_k(x) = (x - x_1) \cdots (x - x_{k-1})(x - x_{k+1}) \cdots (x - x_n)$. The Lagrange interpolating polynomial is then

$$y_1 \frac{q_1(x)}{q_1(x_1)} + y_2 \frac{q_2(x)}{q_2(x_2)} + \cdots + y_n \frac{q_n(x)}{q_n(x_n)}.$$

Even though the Lagrange formula solves our polynomial interpolation problem, it is interesting to look at an alternative formulation. This time, we express the interpolating polynomial $p(x)$ in ascending form $a + bx + cx^2 + dx^3$, where the coefficients are unknowns to be determined. For the graph of p to pass through $(-1, 4)$ we must have $p(-1) = 4$.

This gives the equation

$$a + b(-1) + c(-1)^2 + d(-1)^3 = 4.$$

Combined with the corresponding equations for the other points, this produces a system of four equations in the four unknown coefficients. In matrix form, it is

$$\begin{bmatrix} 1 & -1 & (-1)^2 & (-1)^3 \\ 1 & 2 & 2^2 & 2^3 \\ 1 & 3 & 3^2 & 3^3 \\ 1 & 4 & 4^2 & 4^3 \end{bmatrix} \begin{bmatrix} a \\ b \\ c \\ d \end{bmatrix} = \begin{bmatrix} 4 \\ 5 \\ -2 \\ 1 \end{bmatrix}.$$

(A review of matrix notation and properties is available at the website for this book [87].)

In this formulation we see that polynomial interpolation can be reduced to a system of linear equations. The array that appears on the left of the system is called a *Vandermonde matrix*, a square matrix with rows of the form $[1 \ \ r \ \ r^2 \ \ \ldots \ \ r^{n-1}]$ for n distinct values of r. If we label those values in order $r_1, r_2, \ldots, r_n$, then the ij entry of the Vandermonde matrix is r_i^{j-1}. Polynomial interpolation problems can always be formulated as linear systems with Vandermonde matrices.

Vandermonde matrices are usually mentioned in linear algebra courses in connection with a formula for their determinants, given by the product of all factors $(r_j - r_k)$ with $j > k$. In particular, if V is Vandermonde with distinct r_j's, the determinant of V is nonzero. According to the theory of linear equations, this implies that

$$V \begin{bmatrix} c_1 \\ c_2 \\ \vdots \\ c_n \end{bmatrix} = \begin{bmatrix} y_1 \\ y_2 \\ \vdots \\ y_n \end{bmatrix}$$

has a unique solution for any choice of the y's. In this way, methods of linear algebra prove that there is a unique polynomial of degree $n - 1$ interpolating any n points (x_j, y_j) with distinct x coordinates.

2.6   Palindromials

In Chapter 1 we encountered reverse polynomials. If $p(x) = 3x^2 + 5x + 7$ then its reverse polynomial is $7x^2 + 5x + 3$. In general, $r \neq 0$ is a root of a polynomial p if and only if $1/r$ is a root of the reverse polynomial for p. This follows from the fact that the reverse of $p(x)$ is given by $x^n p(1/x)$ where n is the degree of p.

In this section we consider polynomials that are self-reversing, that is, polynomials whose coefficients form a symmetric pattern, as in $x^5 + 7x^4 - 2x^3 - 2x^2 + 7x + 1$, or equivalently, which satisfy the identity $p(x) = x^n p(1/x)$. We will call them *palindromic polynomials* or *palindromials*. They have a surprising range of applications. Here we will focus on some of their basic properties. In Sidebar 2.1 one application is described. Additional applications and related topics come at the end of the chapter.

An Application of Matrix Palindromials

One of the applications of palindromials arises in mechanical vibrational analysis for physical structures such as buildings, machines, and vehicles. In this context, polynomial eigenvalue problems arise, characterized by an equation

$$P(\lambda)x = 0$$

where $P(\lambda)$ is a polynomial with matrix coefficients. Sometimes P turns out to be a palindromial. With matrix coefficients, $P(\lambda)$ is defined to be a palindromial if its reverse polynomial equals its transpose $P(\lambda)^T$.

This subject is discussed by Mackey et al [113, 114]. One of their examples concerns rail traffic noise caused by high speed trains. It includes a structural model for a rail, as illustrated in Fig. 2.7.

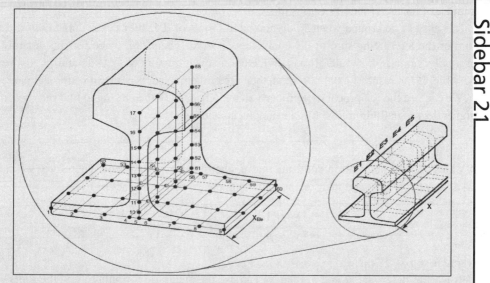

2.7. Structure of a train rail.

The palindromic eigenvalue problem that arises in this example is

$$(\lambda^2 A + \lambda B + A^T)v = 0$$

where A and B are matrices and $B^T = B$. The authors show how the palindromic structure can be used in solving the eigenvalue problem.

Sidebar 2.1

It is convenient to assume that the constant term of a palindromial is nonzero. That is not a significant restriction because we can eliminate a zero constant term through factoring. For example, $0x^4 + 3x^3 - 7x^2 + 3x + 0$ can be rewritten $x(3x^2 - 7x + 3)$. Going a step further, we reach $3x(x^2 - \frac{7}{3}x + 1)$. This illustrates a general result. A self-reversing polynomial can always be expressed $ax^k p(x)$ where $p(x)$ is a monic palindromial. Hereafter we will consider only monic palindromials, all of which have first and last coefficients of 1, and hence only nonzero roots. With this convention, $p(x)$ is a palindromial of degree n if and only if $p(0) = 1$ and $p(x) = x^n p(1/x)$.

If r is a root of a palindromial p, then $1/r$ is a root of the reverse polynomial, which is p itself. This suggests that for palindromials, roots come in reciprocal pairs. However, if $r = \pm 1$, then r is its own reciprocal, and we do not necessarily get a *pair* of roots. For example, $x + 1$ is a palindromial with just one root. This situation occurs for any odd degree palindromial $p(x)$. Since $p(x)$ has an odd number of roots they obviously cannot be paired up. However, an odd degree palindromial always has -1 as a root. We can see this for degree 5 as follows. A fifth degree (monic) palindromial must take the form

$$p(x) = x^5 + ax^4 + bx^3 + bx^2 + ax + 1$$

and $p(-1) = -1 + a - b + b - a + 1 = 0$. The same thing happens for any odd degree. Substituting -1 for x produces the alternating sum of the coefficients, and with an even number of terms, symmetry causes this sum to vanish.

It is easy to determine when a polynomial has roots of ± 1; they occur if and only if the sum or the alternating sum of the coefficients is zero. Therefore, when looking for roots of a polynomial, we should always divide out as many factors of $(x - 1)$ and $(x + 1)$ as possible. If the result is a palindromial, then the remaining roots come in reciprocal pairs.

We can use the reciprocal root property to our advantage when trying to find the roots of higher degree palindromials. As an example, consider

$$p(x) = x^4 + 7x^3 - 2x^2 + 7x + 1.$$

A quick mental calculation confirms that neither 1 nor -1 is a root. Therefore, we can assume roots of r, s, and their reciprocals. This means that $p(x)$ factors as

$$p(x) = (x - r)(x - 1/r)(x - s)(x - 1/s)$$
$$= (x^2 - ux + 1)(x^2 - vx + 1)$$

where $u = r + 1/r$ and $v = s + 1/s$.

If we can determine u and v, then we can use the quadratic formula to find the roots of p. So let us multiply the factors $(x^2 - ux + 1)$ and $(x^2 - vx + 1)$ and match the coefficients with p. That leads to

$$x^4 - (u + v)x^3 + (2 + uv)x^2 - (u + v)x + 1 = x^4 + 7x^3 - 2x^2 + 7x + 1$$

and hence to

$$u + v = -7$$
$$uv = -4,$$

which is readily solved for u and v. This system has a form that will appear frequently later. It specifies the sum (-7) and product (-4) of two unknowns, and translates immediately to the quadratic equation

$$x^2 + 7x - 4 = 0.$$

This has u and v as roots, as we will see in greater detail in Chapter 3.

So u and v are

$$\frac{-7 \pm \sqrt{65}}{2}$$

leading to the factorization

$$p(x) = \left(x^2 + \frac{7 + \sqrt{65}}{2}x + 1\right)\left(x^2 + \frac{7 - \sqrt{65}}{2}x + 1\right).$$

The four roots of p are the roots of the quadratic factors. Applying the quadratic formula to the first factor we obtain roots of

$$\frac{-7 - \sqrt{65} + \sqrt{98 + 14\sqrt{65}}}{4} \quad \text{and} \quad \frac{-7 - \sqrt{65} - \sqrt{98 + 14\sqrt{65}}}{4}.$$

You can check that these roots are reciprocals, as expected.

In this example we have found the roots of a quartic equation. In general solving quartics is more complicated, but with the extra symmetry of a palindromial, everything can be reduced to solving quadratic equations. For higher degree equations, palindromial symmetry implies a similar simplification, as we will shortly see.

Take note of the form in which the roots appear in the preceding example. It is an instance of a radical expression, meaning a combination of numbers using normal operations of arithmetic, as well as radicals. Here only square roots appear, but in a general radical expression there can be cube roots, fourth roots, and so on. For polynomials of degree 4 or less, roots can always be expressed in terms of radicals. This idea will be developed at great length in Chapter 4. For polynomials of degree 5 or more, roots *cannot* always be expressed in terms of radicals. As a result, one does not typically attempt to find exact solutions to polynomials of these degrees. However for palindromials, we can find exact solutions in terms of radicals well beyond the fourth degree.

As an illustration of why this occurs, let us look at a generic sixth degree palindromial,

$$p(x) = x^6 + ax^5 + bx^4 + cx^3 + bx^2 + ax + 1.$$

We know that p has no roots of zero, so $p(x)/x^3$ has the same roots as p. Therefore, the roots of p satisfy

$$x^3 + ax^2 + bx^1 + c + bx^{-1} + ax^{-2} + x^{-3} = 0.$$

Grouping terms leads to

$$(x^3 + 1/x^3) + a(x^2 + 1/x^2) + b(x + 1/x) + c = 0. \tag{4}$$

This looks almost like a polynomial of degree 3, and we can convert it to one by introducing the new variable $u = x + 1/x$. Since

$$(x + 1/x)^2 = x^2 + 2 + 1/x^2,$$

we have

$$x^2 + 1/x^2 = (x + 1/x)^2 - 2 = u^2 - 2.$$

Similarly,

$$(x + 1/x)^3 = x^3 + 3x + 3/x + 1/x^3$$

so

$$x^3 + 1/x^3 = (x + 1/x)^3 - 3(x + 1/x) = u^3 - 3u.$$

Substituting in (4) produces

$$u^3 - 3u + a(u^2 - 2) + bu + c = 0,$$

or

$$u^3 + au^2 + (b - 3)u + c - 2a = 0.$$

In this way, we can reduce a sixth degree palindromial equation to a third degree equation. If we find its roots we can reconstruct the values of x by writing $u = x + 1/x$ in the form $x^2 - ux + 1 = 0$ and applying the quadratic formula. This is what we did in the first example in this section.

The same process can be applied to any even degree palindromial. If $p(x)$ is a palindromial of degree $2n$, then $p(x) = 0$ can be reduced to an equation of degree n in $u = x + 1/x$, because $x^k + 1/x^k$ can always be expressed as a polynomial in u. This can be established by induction. For the induction step, from

$$(x^k + 1/x^k)(x + 1/x) = x^{k+1} + 1/x^{k+1} + x^{k-1} + 1/x^{k-1}$$

we get

$$x^{k+1} + 1/x^{k+1} = (x^k + 1/x^k)(x + 1/x) - (x^{k-1} + 1/x^{k-1}).$$

Thus, if $x^k + 1/x^k$ is given by a polynomial $f(u)$ and $x^{k-1} + 1/x^{k-1}$ is given by a polynomial $g(u)$, then

$$x^{k+1} + 1/x^{k+1} = uf(u) - g(u), \tag{5}$$

and so is a polynomial in u as well. Because this induction step derives the result for $k + 1$ using the two preceding cases, we must independently establish two initial cases. But we have already seen that the result holds for $k = 2$ and $k = 3$.

What about palindromials of odd degree? An odd degree palindromial always has -1 as a root and so is divisible by $x + 1$. Dividing by $x + 1$ leaves us with a factor of even degree. If that is a palindromial then we are back in the earlier case, and can proceed with the $u = x + 1/x$ substitution. In fact, this always occurs.

We can say more. Products and quotients of palindromials are always palindromials. Here, a quotient of polynomials is assumed to have no remainder, so we say $f(x)$ is the quotient $p(x)/g(x)$ just when $p(x) = f(x)g(x)$. In this case, if p and g are both palindromials, then so is f, and if f and g are both palindromials, so is p.

To prove these assertions, let the degree of f be m and the degree of g be n. Then the degree of p is $m + n$. Now assume that both f and g are palindromials. That implies $f(x) = x^m f(1/x)$, $g(x) = x^n g(1/x)$, and $f(0) = g(0) = 1$. Combining these facts, we find

$$\begin{aligned} p(x) &= f(x)g(x) \\ &= x^m f(1/x)x^n g(1/x) \\ &= x^{m+n} p(1/x), \end{aligned}$$

and $p(0) = f(0)g(0) = 1$. Thus p is a palindromial.

For the remaining assertion, assume that p and g are both palindromials. Then since $p(0) = g(0) = 1$ and $p(0) = f(0)g(0)$, we have $f(0) = 1$. Also, we know $p(x) = x^{m+n} p(1/x)$ and $g(x) = x^n g(1/x)$. Therefore,

$$f(x)g(x) = p(x)$$
$$= x^{m+n} p(1/x)$$
$$= x^m f(1/x)x^n g(1/x)$$
$$= x^m f(1/x)g(x).$$

This shows that $f(x) = x^m f(1/x)$. Thus, f is a palindromial.

Returning to our earlier discussion, we now know that an odd degree palindromial $p(x)$ can be divided by $(x+1)$ to produce an even degree palindromial, and then the $u = x + 1/x$ substitution reduces the degree by half. In fact, we can divide out multiple factors of $(x+1)$ and $(x-1)$, where they are present, although this situation rarely arises. Even neglecting this refinement, any palindromial of degree nine or less can be reduced to a polynomial of degree at most four, and thus solved in terms of radicals.

Another Example.   Let us attempt to find the roots of

$$x^9 + 4x^8 - 3x^7 + 6x^6 - x^5 - x^4 + 6x^3 - 3x^2 + 4x + 1 = 0.$$

Because the palindromial has odd degree, we know that -1 is a root. Therefore we can factor out $(x + 1)$ leaving

$$x^8 + 3x^7 - 6x^6 + 12x^5 - 13x^4 + 12x^3 - 6x^2 + 3x + 1 = 0. \tag{6}$$

After checking that there are no remaining roots of 1 or -1 we divide by x^4 and group according to coefficient. Then we have

$$(x^4 + 1/x^4) + 3(x^3 + 1/x^3) - 6(x^2 + 1/x^2) + 12(x + 1/x) - 13 = 0. \tag{7}$$

The next step is to substitute $u = x + 1/x$. We found earlier that $x^2 + 1/x^2 = u^2 - 2$ and $x^3 + 1/x^3 = u^3 - 3u$. Using these with (5) we get $x^4 + 1/x^4 = u^4 - 4u^2 + 2$. Thus substituting into (7) yields

$$u^4 + 3u^3 - 10u^2 + 3u + 1 = 0.$$

And LOOK! It's another palindromial! The substitution $v = u + 1/u$ reduces it to

$$v^2 + 3v - 12 = 0,$$

and that tells us

$$v = \frac{-3 \pm \sqrt{57}}{2}.$$

For each choice of v, we solve for u, expressing the defining equation $v = u + 1/u$ in the form

$$u^2 - vu + 1 = 0.$$

Thus,

$$u = \frac{v \pm \sqrt{v^2 - 4}}{2}.$$

Taking

$$v = \frac{-3 - \sqrt{57}}{2}$$

leads to

$$u = \frac{-3 - \sqrt{57} \pm \sqrt{50 + 6\sqrt{57}}}{4},$$

giving two of the four valid choices of u, and the remaining two u's are found similarly. To determine the roots x, we can solve another quadratic equation, this time

$$x^2 - ux + 1 = 0.$$

For each u there are two solutions, leading to radical expressions for eight roots of the original polynomial. These are complicated so finding them will be left to the interested reader.

This example reveals an entire new vista — *doubly palindromic* polynomials, those for which the degree reducing change of variables produces another palindromial. Examples are easy to obtain. Start with a palindromic polynomial in u, substitute $x + 1/x$ for u, and then multiply by a power of x to eliminate any x's in denominators. Applying this to the arbitrary quartic palindromial $u^4 + au^3 + bu^2 + au + 1$ produces

$$x^8 + ax^7 + (b + 4)x^6 + 4ax^5 + (2b + 7)x^4 + 4ax^3 + (b + 4)x^2 + ax + 1.$$

This is the general doubly palindromic polynomial of degree eight: every eighth degree doubly palindromic polynomial must have this form. When $a = 3$ and $b = -10$ we obtain the example of (6). And we can find the general doubly palindromic polynomial of any specified degree similarly.

Carrying on in this way it is possible to produce triply palindromic polynomials, quadruply palindromic polynomials, and so forth. This gives polynomials of arbitrarily high degree that can be solved using radicals.

Another noteworthy category of palindromials are those of the form

$$p(x) = x^{n-1} + x^{n-2} + \cdots + x + 1,$$

the roots of which are complex n^{th} roots of unity. These are discussed in Appendix A at the website for this book [87].

If $n = 5$ there are five complex fifth roots of unity, 1 and the roots of

$$x^4 + x^3 + x^2 + x + 1 = 0.$$

Using either the factorization method from the start of this section, or the substitution method, the roots are

$$\frac{\sqrt{5} - 1}{4} \pm i\frac{\sqrt{10 + 2\sqrt{5}}}{4} \quad \text{and} \quad \frac{-(\sqrt{5} + 1)}{4} \pm i\frac{\sqrt{10 - 2\sqrt{5}}}{4}.$$

Complete details for the derivation using the factorization method appear in Appendix A.

   These results can also be found geometrically. As shown in Appendix A, one of the fifth roots of unity is $\omega = \cos 72° + i \sin 72°$, and the remaining roots are ω^j for $j = 2, 3, 4, 5$. This shows that the fifth roots of unity can be found once we know exact values for $\cos 72°$ and $\sin 72°$. See Sidebar 2.2 for a geometric construction of these values.

Geometric Derivation of Fifth Roots of Unity

Finding the fifth roots of unity is equivalent to determining expressions for the cosine and sine of $2\pi/5 = 72°$. This can be done geometrically. In Fig. 2.8 are two images of an isoceles triangle ABC with base angles of $72°$ and base AB of length 1. On the right, the triangle has been partitioned into smaller triangles, so that BD and BE are each of length 1. This can be done by striking off an arc of radius 1 with center B. Then all the angles marked with a single arc measure $72°$, all those marked with two arcs are $36°$, and the three smaller triangles are all isosceles. This implies $AD = DE = EC$ and we denote their common length a.

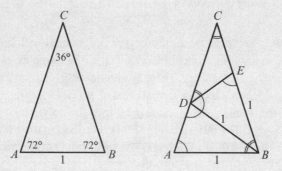

Figure 2.8. Isosceles triangles with base angles of $72°$.

$\triangle ADB$ is similar to $\triangle ABC$ so

$$\frac{1+a}{1} = \frac{1}{a}$$

or $a^2 + a - 1 = 0$. Now we can find a, and hence $\cos 72°$ and $\sin 72°$. We will get

$$\cos 72° = \frac{\sqrt{5}-1}{4} \quad \text{and} \quad \sin 72° = \frac{\sqrt{10 + 2\sqrt{5}}}{4}.$$

Sidebar 2.2

2.7   Marden's Theorem

The last topic for this chapter is a personal favorite, a result I call Marden's theorem. It establishes a fantastic relation between the geometry of plane figures and the related root locations for a polynomial and its derivative. Here the goal will be to understand the statement of the theorem. A more complete discussion and a proof can be found at the website for this book [87].

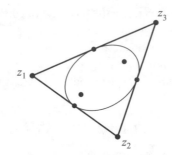

Figure 2.9. Configuration for Marden's theorem.

Although the theorem has a more general statement, the version I like best involves a cubic polynomial $p(z)$, whose roots are the vertices of a triangle in the complex plane. It is known that the roots of $p'(z)$ must lie somewhere within the triangle. That is a consequence of Lucas's theorem, which says that the roots of the derivative of p must lie in the smallest convex polygon containing the roots of p. When the polygon is a triangle and p is a cubic, Marden's theorem says that the roots of the derivative lie at the foci of an inscribed ellipse.

Marden's Theorem.   Let $p(z)$ be a third-degree polynomial with complex coefficients whose roots z_1, z_2, and z_3 are noncollinear points in the complex plane. Let T be the triangle with vertices at z_1, z_2, and z_3. There is a unique ellipse inscribed in T and tangent to the sides at their midpoints. Its foci are the roots of $p'(z)$.

If the vertices of the triangle are z_1, z_2, and z_3, then the polynomial p may be taken as $(z - z_1)(z - z_2)(z - z_3)$. The ellipse in Fig. 2.9 is inscribed in the triangle and is tangent to the sides of the triangle at their midpoints. The foci of the ellipse, marked as black dots, are the roots of $p'(z)$.

The website for this book [87] provides a link to a web based treatment of Marden's theorem, hosted by the on-line mathematics journal JOMA [89]. This comprehensive presentation offers a complete proof and extensive background information, as well as historical references and animated graphics illustrating the theorem. In one animation, three ellipses share a common pair of foci. These are symmetric with respect to the centroid of the triangle whose vertices are the roots of $p(z)$. Each ellipse passes through the midpoint of one side of the triangle. The configuration evolves dynamically as the foci move about in the

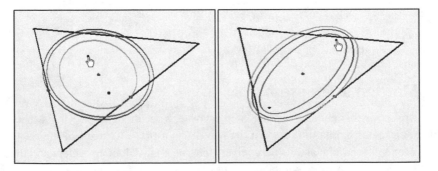

Figure 2.10. Screen images from animated graphic for Marden's theorem.

plane. As the foci approach the roots of the derivative, the three ellipses coalesce into one, inscribed in the triangle and tangent at the midpoints of the sides. Two images from this animation appear in Fig. 2.10.

2.8   Solution to a Perplexing Puzzle

In the puzzle, an unknown polynomial $p(x)$ has whole number coefficients. One character, the pussycat, finds out the values of $p(1)$ and $p(p(1))$, and then is able to completely determine the polynomial. Here is the explanation.

Suppose that the polynomial is $p(x) = a_n x^n + \cdots + a_0$. The first clue, $p(1)$, gives the sum of all the coefficients. Because they are whole numbers, each coefficient must therefore be less than $p(1)$, which we now denote b. The pussycat next asks for the value of $p(b)$. Since each coefficient is less than b, $p(b) = a_n b^n + \cdots + b_0$ is the base b expansion of the result, with the coefficients of p appearing as the digits. Thus, by expressing $p(p(1))$ in base b, the coefficients of p are obtained.

This same strategy works for any base greater than or equal to $p(1)$. One base that is particularly handy is 10^k, where k is the smallest integer such that $10^k \geq p(1)$. The value of $p(10^k)$ is then a base 10^k expansion,

$$a_n 10^{nk} + a_{n-1} 10^{(n-1)k} + \cdots + a_1 10^k + a_0.$$

Each a_k is less than 10^k and so has at most k digits. These coefficients may be read off in blocks of size k in the value of $p(10^k)$.

For example, suppose that the polynomial is $p(x) = 6x^5 + 22x^4 + 7x^3 + 11x^2 + 21x + 8$. Then $p(1) = 75$ so we take $k = 2$. Now we are given the value of

$$p(100) = 6 \cdot 10^{10} + 22 \cdot 10^8 + 7 \cdot 10^6 + 11 \cdot 10^4 + 21 \cdot 10^2 + 8$$
$$= 62207112108.$$

We read the coefficients off this result in two digit blocks: $06, 22, 07, 11, 21, 08$.

2.9   History, References, and Additional Reading

The perplexing puzzle has appeared in various places with different attributions. One correspondent reported that it appeared in a 1960's problem collection at a school in Moscow (then in the USSR), and probably had appeared earlier. It also appeared in 2005 in the *College Math Journal* [95], with an attribution to an earlier publication. In a follow-up note, Bornemann and Wagon [21] show that it is possible to determine the polynomial from the value of just one $p(x)$.

The derivation of formulas for the maximum and minimum of two numbers using the quadratic formula was published previously in [76].

As mentioned earlier, the solution of cubic equations using the curly root function was based on papers of Nogrady [123] and Pettit [130]. Traditionally, solutions to polynomial equations have been formulated in terms of radicals, with regard to which much more will be said in Chapter 4. Quintic equations are not generally solvable using radicals, but their solutions can be expressed in terms of elliptic functions (see [96]). For exponential-linear equations (those having the form $a^x = bx + c$) an approach like the curly root solution of cubics will be presented in Chapter 9.

The material on chaos in Newton's method is adapted from the paper by Saari and Urenko [139]. Some general treatments of chaos are provided by [38, 39, 61]. For an elementary discussion of basic concepts of chaos in difference equation models, see [78, Chapter 14].

There is a vast literature on polynomial interpolation. Introductory comments on the subject can be found in [8, Chapter 7], [23, pp. 8–10], and [132, pp. 133–139]. For a discussion of Vandermonde matrices, including their connection with interpolation, see [75].

Palindromials are more commonly referred to in print as reverse or reciprocal polynomials. They have a long history of inclusion in algebra textbooks dating at least to 1811 [72]. Today, palindromials appear in many contexts and have many applications. For example, they arise in knot theory in connection with polynomial invariants of knots with mirror image symmetry. There is also interest in characterizing palindromials over finite fields, with applications to reversible cyclic codes, high minimum distance BCH codes, and efficient finite field arithmetic.

Here is a sample of accessible articles on palindromials. An application in plane geometry is presented by Clarke [31]. This work concerns sequences of polygons in which each term is formed by connecting the midpoints of the sides of its predecessor. Convergence of such sequences is shown to be related to certain palindromic polynomials. Quadratic palindromials are used to derive the quadratic formula by Spiegel [151]. Viana and Veloso [162] describe Galois theory in the context of palindromials. Finally, Eaton [43] uses palindromials and facts about determinants to prove the Fundamental Theorem of Algebra, a result we will consider in Chapter 4.

As mentioned, more information about Marden's theorem appears online in [89], a link to which can be found at the website for this book [87]. An additional reference is [88].

3

Polynomial Roots and Coefficients

The most common standard form for a polynomial is descending form. Given a polynomial in any form whatever, routine algebra readily produces the polynomial's descending form. The same cannot be said about factored form. For example, if $p(x) = 3x^5 + 15x^4 + 30x^3 + 30x^2 - 7$, it is not obvious how to find the factored form, or even if one exists. But if the factored form is given, there is no difficulty finding the descending form. Perhaps it is for this reason that the descending form is usually considered the default representation of a polynomial. And since the descending form is completely determined by the sequence of coefficients, they give one of the most fundamental descriptions of a polynomial. From this point of view, the coefficients are rather like DNA for the polynomial.

On the other hand, often we would like to solve a polynomial equation. Such an equation can always be put into the form $p(x) = 0$, in which case the solutions are the roots of the polynomial. The Root-count theorem (page 5) tells us that there can be at most n roots for a polynomial of degree n, and in the best of possible cases, all n roots can be found. Then the polynomial can be written in factored form

$$p(x) = A(x - r_1)(x - r_2) \cdots (x - r_n).$$

Here, the n roots, $r_1, r_2, \ldots, r_n$ (together with the leading coefficient A) carry complete information about the polynomial. So the roots (plus the leading coefficient) can be thought of as an alternate representation of the polynomial's DNA. Relationships between these two representations, given by the roots and the coefficients, is the focus for this chapter.

Unless noted otherwise, we shall assume henceforth that our polynomials are *monic*: $x^n + a_{n-1}x^{n-1} + \cdots + a_0$. In this case, the factored form requires no leading coefficient (as $A = 1$) and it is correct to say that the polynomial is completely determined by its n roots.

Given the roots, we can easily find the coefficients. Far more interesting and mathematically significant is the reverse process, finding the roots given the coefficients. This is the essence of solving a polynomial equation, and is the focus for the next chapter. But even without finding the roots explicitly, it is possible to uncover many facts about them. For example, by simply looking at a polynomial in descending form, one can immediately deduce the values of the sum of the roots, the product of the roots, even the sum of the reciprocals of the roots. These are special cases of a more general phenomenon involving symmetric combinations of the roots.

Exploring the relations between the roots of a polynomial and its coefficients leads naturally to a number of interesting concepts. Some highlights are the following:

- Symmetric functions

- Reciprocal roots

- Reverse polynomials

- Discriminants

- Newton's identities.

The ideas developed here have fundamental significance to the problem of solving polynomial equations, and will be featured in Chapter 4. But they are also interesting in their own right, and deserve to be better known.

3.1   Roots and symmetric functions

First we consider how the coefficients of a polynomial depend on the roots. Suppose, for example, that the roots of a polynomial are 2, 3, 5, and -11. Then we can immediately write

$$p(x) = (x - 2)(x - 3)(x - 5)(x + 11)$$

and then obtain the descending form

$$p(x) = x^4 + x^3 - 79x^2 + 311x - 330.$$

This illustrates a simple but important principle: the coefficients are determined by the roots.

As is often the case, working with specific numbers as in the preceding example obscures important patterns. Repeating the same process with roots a, b, c, and d we find that

$$p(x) = (x - a)(x - b)(x - c)(x - d) \tag{1}$$

leads to

$$p(x) = x^4 - (a + b + c + d)x^3 + (ab + ac + ad + bc + bd + cd)x^2$$
$$- (abc + acd + abd + bcd)x + abcd. \tag{2}$$

Now there are obvious patterns. The constant coefficient is the product of the roots, and the coefficient of x^3 is the negative of the sum of the roots.   Neglecting signs, all the coefficients are sums of products of roots. For the x^2 term the products have all possible combinations of two roots, for the x term all combinations of three roots.

A moment's reflection reveals why this pattern occurs. Transforming the factored form of (1) to the descending form in (2) amounts to repeated application of the distributive law.

Each term in the descending form in (2) is made up of four factors, one from each set of parentheses on the right side of (1). We create a term for every possible way of selecting one factor from each parenthesis. But for each parenthesis we have two choices, either x or the negative of one of the roots. Out of these combinations, which terms involve an x^2? These result from selecting x from two of the parentheses, and negatives of roots from the other two. Putting them all together shows that the x^2 term must have a coefficient that is the sum of all possible products of two elements from the set $\{-a, -b, -c, -d\}$. Extending this analysis to the general case, the coefficient of x^k must be made up of all possible products of $n - k$ of those elements.

This leads in a natural way to what are called *elementary symmetric functions*. Given a set of n elements, the sum of all products of k of the elements is called the the kth elementary symmetric function, denoted σ_k. Thus, for example,

$$\sigma_3(a, b, c, d) = abc + abd + acd + bcd,$$

$$\sigma_2(a, b, c) = ab + ac + bc,$$

and

$$\sigma_1(a, b, c, d, e) = a + b + c + d + e.$$

In general, the coefficients of a polynomial are symmetric functions of the negatives of the roots. For even k, σ_k produces the same result for the negatives of the roots as it does for the roots themselves: multiplying an even number of factors causes the negative signs to cancel out. For odd k, applying σ_k to the negatives of the roots produces the negative of what you get for the roots themselves. Thus

$$p(x) = (x - r_1)(x - r_2) \cdots (x - r_n)$$

appears in descending form as

$$p(x) = x^n + a_{n-1}x^{n-1} + \cdots + a_0$$

where

$$a_k = (-1)^{n-k} \sigma_{n-k}(r_1, r_2, \ldots, r_n).$$

The reference to symmetry in the name of σ_k reflects the fact that permuting the order of the roots has no effect on the value of σ_k. So, for example, $\sigma_k(r_1, r_2, r_3) = \sigma_k(r_2, r_1, r_3) = \sigma_k(r_3, r_2, r_1)$. This amounts to the observation that the coefficients of a polynomial are not changed by rearranging the order of the factors $(x - r_j)$.

Perhaps you are wondering what that has to with do with symmetry. The answer reflects an abstract concept of symmetry that mathematicians have adopted. In everyday usage, a figure has symmetry if it looks the same as its mirror image. Put another way, if you operate on the figure in a certain way (by forming its reflection), there is no observable difference. We can extend this notion of symmetry to other sorts of operations. If you rotate a five pointed star through a fifth of a full revolution, the result is indistinguishable from the starting figure. By analogy, that is another kind of symmetry. We might say that one figure has mirror symmetry and another has rotational symmetry.

As used in mathematics, *symmetry* can refer to any sort of operation at all. We can say that e^x is symmetric with respect to differentiation, for example. The σ's are symmetric

Sidebar 3.1

Mathematical Symmetry

Everyone is familiar with the concept of symmetry, and yet it is surprisingly difficult to define concisely and completely. As explained in the main narrative, in the modern mathematical view point symmetry is defined in terms of invariance under some kind of transformation or collection of transformations. After some consideration, this idea can come to appear quite natural and self evident. But it is not the sort of thing that first springs to mind when most people try to define symmetry.

Where and when did the mathematical idea of symmetry arise? According to Stewart, the initial impetus was work around 1830 of the French mathematician Galois, who analyzed solvability of polynomial equations in connection with permutations of the roots. Stewart summarizes this development in a discussion of the question "What is symmetry?" He says

> Before Galois, all answers to this question were rather vague, handwavy things, with appeals to features like elegance of proportion.... After Galois — and after a short period during which the world of mathematics sorted out the general ideas behind his very specific application — there was a simple and unequivocal answer.... A symmetry of some mathematical object is a transformation that preserves the object's structure [153, p. 118].

This idea has proven to have profound consequences in mathematics and science. In physics, symmetries are connected with conservation laws by Noethe's theorem, and provide a unifying viewpoint throughout the subject [108]. In mathematics, Galois' work led to Felix Klein's formulation in 1872 of a new conception of geometry in which the transformational idea of symmetry is fundamental. Klein's formulation, called the Erlangen Program, recognizes many different kinds of geometry, each associated with a particular set of transformations. Quoting Stewart, again,

> For example, the symmetries of Euclidean geometry are those transformations of the plane that preserve lengths, angles, lines, and circles. This is the group of

with respect to reorderings or *permutations* of the variables. This kind of symmetry has a geometric counterpart as well. A function $f(x, y)$ is called symmetric if it is unchanged by swapping x and y. In this case, for any constant k the graph of $f(x, y) = k$ has mirror symmetry in the line $x = y$.

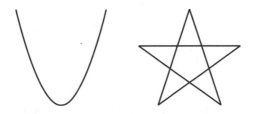

Figure 3.1. A parabola has mirror symmetry; a star has rotational symmetry.

Mathematical Symmetry (cont.)

all rigid motions of the plane. Conversely, anything that is invariant under rigid motions falls naturally within the purview of Euclidean geometry [153, p. 160].

Adopting this point of view has two consequences. First, there is a transition of how we think of geometry and symmetry. Symmetry is not just a property that certain geometric figures can possess, but becomes the central concern of the entire subject. Second, this more general viewpoint allows us to consider other subjects as analogous to plane geometry in a natural way. This guides intuition, permits techniques developed in one setting to be adapted for use in other settings, and contributes to cross fertilization among different branches of mathematics.

It is ironic that today, one's first exposure to the term symmetry in connection with roots of equations seems a bit obscure. After seeing the explanation, we can understand this idea as an extension of the more geometrical notions of symmetry we all grow up with. Now we can complete the circle. The extended idea of symmetry arising out of Galois' work with roots of equations has become the founding concept of geometry itself.

Galois   Klein

Sidebar 3.1 (cont.)

The effects of permuting subsets of roots played a fundamental role in the evolution of our understanding of polynomials. They led to the development of group theory and Galois theory, standard topics in college courses in abstract algebra. It is in this historical context that elementary symmetric functions assume their greatest significance. But even at the more down-to-earth level of this book, the symmetric functions reveal interesting things about roots of polynomials, as we shall see.

It might seem that knowing how to go from roots to coefficients should make it easy to solve polynomial equations. After all, solving a polynomial equation means finding the roots given the coefficients. Why can't we just reverse the process of finding the coefficients from the roots?

To make this question concrete, consider the polynomial

$$p(x) = x^3 + 5x^2 - 6x + 3,$$

and the problem of determining its roots. That amounts to finding a, b, and c so

$$p(x) = (x - a)(x - b)(x - c).$$

We know that a, b, and c have to satisfy the equations

$$a + b + c = -5$$
$$ab + ac + bc = -6$$
$$abc = -3.$$

They don't look very complicated. Surely we can just solve them for a, b, and c.

Alas, that task is harder than it appears. If you have never thought about solving polynomials in this way before, I recommend that you take a little time right now to do so. Try your hand at solving the system of equations above for a, b, and c.

While this method of solving polynomials is difficult, it is not impossible, at least not for polynomials of degree 4 or less. In fact, systems like the one above can be solved for general cubic and quartic equations with an appropriate simple change of variables. It takes some insight to find the right change of variables, as you might expect. We will come to that in Chapter 4.

Even though it is not easy to find roots of polynomials by using elementary symmetric functions, we can make certain observations about *combinations* of roots. Here are a few simple examples.

Sum and Product of the Roots.   The simplest elementary symmetric functions are the first and last. They lead immediately to two observations: the sum of the roots of a polynomial $p(x) = x^n + a_{n-1}x^{n-1} + \cdots + a_0$ is given by $-a_{n-1}$ and the product of the roots by $(-1)^n a_0$. For a quadratic, these are the only two coefficients that matter. This gives us a backdoor solution of a quadratic equation. If the sum of two numbers is A and the product of the two numbers is B, then the numbers are the roots of the quadratic $x^2 - Ax + B$. We will see later that this sort of reasoning can be useful in solving higher degree polynomials.

Average of the Roots.   Knowing the sum of the roots leads to other useful observations. For example, given an arbitrary polynomial, we can easily read off the average of the roots (the sum of the roots divided by the degree). Try it on this one: for $x^6 - 12x^5 + 3x^4 - 2x^3 + 11x^2 - 7x + 5$, the average of the roots is 2. Or, if we know that a polynomial has all real roots, and if the coefficient of the second-highest power of x is negative, then there has to be at least one negative root.

But we can say more. The roots of the derivative of this polynomial have exactly the same average. For the particular case at hand, you can take the derivative, divide by its leading coefficient of 6, and verify that the average of the roots is still 2. For the general case, say the original polynomial is $p(x) = x^n + bx^{n-1} + \cdots$. The average of the roots is $-b/n$. Now look at the derivative: $p'(x) = nx^{n-1} + (n-1)bx^{n-2} + \cdots$. Because our interpretations of the coefficients are only valid for monic polynomials, we have to divide out the leading n. That gives $\frac{1}{n}p'(x) = x^{n-1} + \frac{n-1}{n}bx^{n-2} + \cdots$, which of course has the same roots as p'. What is the average of those roots? Answer: $-b/n$. Thus, we have shown that the average of the roots of a polynomial is the same as the average of the roots of the derivative.

This fact has an interesting implication in the context of Marden's theorem (page 41). Recall that in Marden's theorem, a polynomial p of degree three has roots at the vertices of a triangle in the complex plane. The average of these roots is the centroid of the triangle. Marden's theorem says that the roots of p' are the foci of an inscribed ellipse. Their average is therefore at the center of the ellipse. Thus, since the average of the roots of p is the same as the average of the roots of p', the inscribed ellipse must be centered at the centroid of the triangle.

Shifted Polynomials. Here is one more example. Consider a polynomial $p(x)$ with (unknown) roots $r_1, r_2, \ldots, r_n$. Let $q(x) = p(x-1)$. Clearly this has roots of r_k+1. Generally, replacing x by $x - h$ in a polynomial increases each of the roots by h. Now what will the sum of the shifted roots be? Since each individual root increases by h, the sum increases by nh. Thus, if in the original polynomial the coefficient of the term of degree $n - 1$ is a_{n-1}, then in the shifted polynomial the corresponding coefficient will be $a_{n-1} - nh$. We can make this coefficient vanish by taking $h = a_{n-1}/n$. This is a standard first step in solving a polynomial equation: shifting the variable to eliminate the term of second highest degree. For the cubic, for example, we can consider only polynomials of the form $x^3 + ax + b$, since any cubic can be reduced to one of these. Recall that this fact was used in solving cubic equations with curlyroots (see page 26). A polynomial of degree n with $a_{n-1} = 0$ is called *reduced*. For example, $x^3 + ax + b$ is called a *reduced cubic*.

I hope that these examples have made my point: there is value in knowing how the roots and coefficients of a polynomial are related. Even more, I hope the examples have helped to whet your appetite for the next section of the chapter, in which we will explore symmetric functions and roots in greater depth. Before proceeding, though, we should take a moment to consider the situation for non-monic polynomials. While we can always divide by the leading coefficient to make a polynomial monic, there are times when it is just as convenient to modify our interpretations of the coefficients. For example, for a polynomial with leading coefficient a_n (not necessarily one), the sum of the roots is given by the ratio $-a_{n-1}/a_n$ and the product of the roots is given by a_0/a_n. More generally, the expression $(-1)^{n-k}\sigma_{n-k}(r_1, r_2, \ldots, r_n)$ is equal to the ratio a_k/a_n. Each of these reduces to the corresponding result in the monic case when $a_n = 1$.

3.2   Root Identities

Consider a polynomial given in both factored and descending forms by

$$p(x) = (x - r_1)(x - r_2)\cdots(x - r_n) = x^n + a_{n-1}x^{n-1} + \cdots + a_0.$$

We know $a_k = (-1)^{n-k}\sigma_{n-k}(r_1, \ldots, r_n)$. Because each σ is a symmetric function of the roots, combinations of the σ's are also symmetric functions. Often these combinations can take on very simple forms.

For example, take $n = 3$ and consider $s = \sigma_1^2 - 2\sigma_2$. Direct calculation shows

$$\begin{aligned}
\sigma_1^2 - 2\sigma_2 &= (r_1 + r_2 + r_3)^2 - 2(r_1 r_2 + r_1 r_3 + r_2 r_3) \\
&= (r_1^2 + r_2^2 + r_3^2 + 2r_1 r_2 + 2r_1 r_3 + 2r_2 r_3) - 2(r_1 r_2 + r_1 r_3 + r_2 r_3) \\
&= r_1^2 + r_2^2 + r_3^2.
\end{aligned}$$

The final form is a particularly attractive and simple symmetric function of three variables. It's expression as a combination of elementary symmetric functions raises obvious questions. Does something similar happen for sums of cubes? For sums of kth powers? For arbitrary symmetric functions? One generalization is immediate. For any n, the sum of squares $r_1^2 + r_2^2 + \cdots + r_n^2$ is given by $\sigma_1^2 - 2\sigma_2$.

Although this is a general result about symmetric functions, for polynomials the conclusion becomes an identity linking roots and coefficients:

$$r_1^2 + r_2^2 + \cdots + r_n^2 = a_{n-1}^2 - 2a_{n-2}. \tag{3}$$

Let us apply this to the cubic polynomial $x^3 - 4x^2 + 5x + 3$. Although we do not know the roots of this polynomial, we can conclude that their squares add up to $4^2 - 2 \cdot 5 = 6$. From the previous section, we also know that the sum of the roots is 4 and the product of the roots is -3. It seems that we have a good deal of information about these unknown roots.

It is only natural to wonder whether we have enough information now to discover what the roots are. Can you solve this system?

$$r_1 + r_2 + r_3 = 4$$
$$r_1^2 + r_2^2 + r_3^2 = 6 \tag{4}$$
$$r_1 r_2 r_3 = -3.$$

Give it a try.

How did you make out? Most likely, you did not find a simple way to solve the system, although such a solution method might exist (and would be worth publishing if found). Even if you did not succeed, in all likelihood you made some progress. My own unsuccessful attempts definitely left me feeling I was getting closer to a solution.

But why stop with the equations in system (4)? Perhaps we can find other identities involving the roots, just as we found (3). What about the sum of the cubes of the roots? That is a symmetric function of the roots. Can it be expressed in terms of the coefficients? What about the sum of the fourth powers of the roots?

Pursuing these questions, we can find many identities for symmetric functions of the roots. In discovering them we are following in the footsteps of such luminaries as Isaac Newton. Newton found a great many identities, including these for the three roots of a cubic:

$$r_1^3 + r_2^3 + r_3^3 = -a_2^3 + 3a_2 a_1 - 3a_0$$
$$r_1^2 r_2^2 + r_1^2 r_3^2 + r_2^2 r_3^2 = a_1^2 - 2a_2 a_0$$
$$r_1^2 r_2 + r_1^2 r_3 + r_2^2 r_1 + r_2^2 r_3 + r_3^2 r_1 + r_3^2 r_2 = -a_2 a_1 + 3a_0.$$

He also found a general scheme that produces identities for

$$r_1^k + r_2^k + \cdots + r_n^k$$

for any power k and any number of variables n. More about these identities will be presented below.

General and Elementary Symmetric Polynomials

The *Fundamental Theorem of Symmetric Polynomials* is, as its name suggests, a key result in the subject of root identities. The theorem states that any symmetric polynomial can be represented in terms of the *elementary* symmetric polynomials. For example, suppose r, s, and t are the three roots of a cubic polynomial $p(x) = x^3 + a_2 x^2 + a_1 x + a_0$. Then $r^3 + s^3 + t^3$ is a symmetric polynomial in the roots, but it is not one of the *elementary* symmetric polynomials. They are $\sigma_1 = r + s + t = -a_2$, $\sigma_2 = rs + rt + st = a_1$, and $\sigma_3 = rst = -a_0$.

The fundamental theorem asserts that $r^3 + s^3 + t^3$ is expressible in terms of the σ's, or, equivalently, in terms of the coefficients of p. This assertion is realized in Newton's result

$$r^3 + s^3 + t^3 = \sigma_1^3 - 3\sigma_1\sigma_2 + 3\sigma_3 = -a_2^3 + 3a_2 a_1 - 3a_0.$$

As reported by Edwards [47], Newton had made an extensive study of symmetric root polynomials as early as the mid-1660's, and was very likely aware of the fundamental theorem at that time. Edwards adds "It must be admitted, however, that neither a careful statement nor a proof of it seems to have been published before the nineteenth century. Everyone seemed familiar with it and used it without inhibition."

Newton

Kline [101, p. 600] credits Vandermonde with the first published proof of the fundamental theorem in 1771. However, it should be noted that Vandermonde's version of the result was stated in terms of roots and coefficients of a polynomial. Edwards makes the point that the fundamental theorem is properly about symmetric polynomials in n variables, independent of the context of coefficients and roots of a polynomial. Such a formulation sidesteps any philosophical issues concerning the existence or nature of roots. It is apparently a formulation in this context that Edwards refers to in the quotation above.

Sidebar 3.2

In terms of the foregoing comments, how do things now stand? We are interested in identities that relate roots of a polynomial to its coefficients. The coefficients, themselves, are given by the elementary symmetric functions σ_k. We have also seen that combinations of the σ's are again symmetric functions. And we have seen that a great many symmetric functions can be expressed in terms of the σ's.

At this point it is a reasonable conjecture that *all* symmetric functions can be expressed in terms of the σ's. As a matter of fact, they can (see Sidebar (3.2)). To be more precise, any symmetric *polynomial* of the roots can be expressed as a polynomial combination of the σ's. The upshot is this: because we know the coefficients of a given polynomial, we know the values of all of the σ's. That implies that we can determine any symmetric polynomial of the roots. This in turn may prove to be a powerful tool for finding the roots themselves.

In the next chapter, we will put these ideas to good use, deriving a method for solving quartic equations. As you will see, the approach is somewhat indirect, and relies on the property that reordering the roots leaves symmetric functions unchanged.

But before moving on to solution methods, we should look a bit more closely at identities for roots. First we will look at some particular identities involving differences and reciprocals of the roots. Then we will see an entire system of identities referred to collectively as Newton's identities.

The Discriminant. Readers of this book will be familiar with the *discriminant* $D = b^2 - 4ac$ of the quadratic $ax^2 + bx + c$. As we know, the sign of D determines whether the polynomial has real roots and $D = 0$ indicates a double root. This repeated roots criterion can be extended to polynomials of degree n, using a particular symmetric function of the roots.

Define $D = (r_1 - r_2)^2 (r_1 - r_3)^2 \cdots (r_{n-1} - r_n)^2$, where the product includes one factor $(r_j - r_k)^2$ for each pair of roots. Clearly, D is a symmetric function of the roots, and vanishes precisely when at least two of the roots are equal. Since D is a symmetric function it is a combination of elementary symmetric functions. Thus, for every n there is a discriminant formula in terms of the coefficients. As with quadratics, it is possible to determine whether a polynomial possesses repeated roots without knowing what they are.

Although the discriminant formula gets complicated as the n increases, it is instructive to consider the cases for $n = 2, 3$.

For $n = 2$ we have $D = (r_1 - r_2)^2 = (r_1 + r_2)^2 - 4r_1 r_2$. If we express the quadratic polynomial in the form $x^2 + bx + c$, we know $b = -(r_1 + r_2)$ and $c = r_1 r_2$. Thus, $D = b^2 - 4c$, the familiar form when $a = 1$. We can generalize to the nonmonic case by adjusting as described on page 51. That leads to $D = (b^2 - 4ac)/a^2$. This is not quite the traditional form of the quadratic discriminant, but we can see how the two different versions of the discriminant are related.

For $n = 3$ the algebra is already much more complicated:

$$D = (r_1 - r_2)^2 (r_1 - r_3)^2 (r_2 - r_3)^2.$$

This can be expressed as

$$D = a_1^2 a_2^2 - 4a_2^3 a_0 + 18a_0 a_1 a_2 - 4a_1^3 - 27a_0^2, \tag{5}$$

which is easier to verify than it is to discover. Things are much simpler for a reduced cubic, with $a_2 = 0$. (We saw on page 51 that this can always be arranged by making a simple substitution for the variable.) When $a_2 = 0$ (5) becomes $D = -4a_1^3 - 27a_0^2$. It is more challenging to derive the simplified equation directly from the definition of D. The ambitious reader may wish to attempt this by hand, although it is advisable to delay this effort until after reading about Newton's identities below. With a computer algebra system it is possible to derive an identity like (5) by trial and error (see Sidebar (3.3)).

Reciprocal Roots. So far we have considered only polynomial combinations of the roots. But we can also formulate symmetric functions that are not polynomials. For example, the

---

Using a Computer Algebra System

Computer algebra software, such as Mathematica and Maple, can be used to search for identities in an exploratory fashion. The figure below shows a screen image from a Maple session of such a search. The goal was to express the discriminant of a cubic polynomial in terms of the coefficients.

Treat the three roots of the cubic, r, s, and t, as unknown variables. Define the discriminant as $D = (r-s)^2(r-t)^2(s-t)^2$, and the coefficients as $a_2 = r+s+t$, $a_1 = rs+rt+st$, and $a_0 = rst$. When D is completely expanded its terms are of sixth degree, with no one variable having degree higher than 4. We should take this into account in searching for an equivalent combination of coefficients. For example, terms of $a_1^2 a_2^2$ will have degree 6, again with no variable having degree higher than 4. By trial and error, subtract multiples of appropriate combinations of a's from D, seeking to reduce the result as much as possible. Two steps of this process are shown in the figure. A complete solution means finding a combination of a's that results in zero when subtracted from D. With Maple doing all the work, this is a feasible (even enjoyable) exercise.

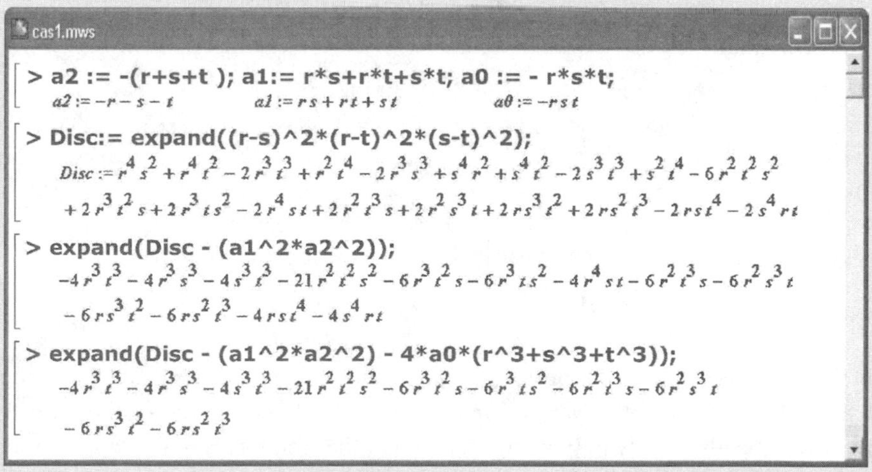

```
cas1.mws
> a2 := -(r+s+t); a1:= r*s+r*t+s*t; a0 := - r*s*t;
    a2 := -r - s - t        a1 := rs + rt + st        a0 := -rst
> Disc:= expand((r-s)^2*(r-t)^2*(s-t)^2);
    Disc := r^4 s^2 + r^4 t^2 - 2 r^3 t^3 + r^2 t^4 - 2 r^3 s^3 + s^2 r^4 + s^4 t^2 - 2 s^3 t^3 + s^2 t^4 - 6 r^2 t^2 s^2
    + 2 r^3 t^2 s + 2 r^3 t s^2 - 2 r^4 s t + 2 r^2 t^3 s + 2 r^2 s^3 t + 2 r s^3 t^2 + 2 r s^2 t^3 - 2 r s t^4 - 2 s^4 r t
> expand(Disc - (a1^2*a2^2));
    -4 r^3 t^3 - 4 r^3 s^3 - 4 s^3 t^3 - 21 r^2 t^2 s^2 - 6 r^3 t^2 s - 6 r^3 t s^2 - 4 r^4 s t - 6 r^2 t^3 s - 6 r^2 s^3 t
    - 6 r s^3 t^2 - 6 r s^2 t^3 - 4 r s t^4 - 4 s^4 r t
> expand(Disc - (a1^2*a2^2) - 4*a0*(r^3+s^3+t^3));
    -4 r^3 t^3 - 4 r^3 s^3 - 4 s^3 t^3 - 21 r^2 t^2 s^2 - 6 r^3 t^2 s - 6 r^3 t s^2 - 6 r^2 t^3 s - 6 r^2 s^3 t
    - 6 r s^3 t^2 - 6 r s^2 t^3
```

Figure 3.2. Screen image of a Maple session.

Sidebar 3.3

expression

$$\frac{1}{r_1} + \frac{1}{r_2} + \frac{1}{r_3} + \cdots + \frac{1}{r_n}$$

is unaffected by reordering the r's, and so is a symmetric function of the roots. But it is not a symmetric *polynomial* in the roots. One might not expect this sort of function to be representable as a combination of the elementary symmetric functions. It may be surprising, therefore, that symmetric functions of the reciprocals of the roots are simply related to the coefficients of the original polynomial.

The key idea involves the *reverse* or *reciprocal* polynomial for a given polynomial $p(x)$, mentioned in the two preceding chapters (see pages 18 and 34). To review, the reverse of $p(x)$ is what we get by reversing the order of the coefficients in p. Thus, if $p(x) = 5x^3 - 8x^2 - 2x + 17$, then the reverse polynomial is $\text{rev}\, p(x) = 17x^3 - 2x^2 - 8x + 5$. Algebraically, the two polynomials are related by

$$\text{rev}\, p(x) = x^n p(1/x),$$

which shows that a nonzero r is a root of $\text{rev}\, p$ if and only if $1/r$ is a root of p. Thus, if p has a nonzero constant term, then the reciprocals of the roots of p are precisely the roots of $\text{rev}\, p$.

Let's apply this to the sum of the reciprocals of the roots of $p(x) = 5x^3 - 8x^2 - 2x + 17$. That is the same as the sum of the roots of $\text{rev}\, p = 17x^3 - 2x^2 - 8x + 5$, and we know from earlier work that this will be $2/17$. (Here, we are adjusting for the fact that $\text{rev}\, p$ is not monic, having a leading coefficient of 17.) In retrospect, we need not have taken the trouble to write down $\text{rev}\, p$. We simply apply the result for the sum of the roots of the original p, but reverse the coefficients.

The same process can be applied to any polynomial $p(x) = x^n + a_{n-1}x^{n-1} + \cdots + a_0$, showing that the sum of reciprocal roots is $-a_1/a_0$. More generally, any identity relating the roots of p and the coefficients a_k can be applied to the reciprocals of the roots by reversing the order of the a_k, and adjusting, if necessary, for a_0 different from 1. For example, we saw earlier that

$$r_1^2 + r_2^2 + \cdots + r_n^2 = a_{n-1}^2 - 2a_{n-2}.$$

Reversing the coefficients now gives us

$$\frac{1}{r_1^2} + \frac{1}{r_2^2} + \cdots + \frac{1}{r_n^2} = \frac{a_1^2}{a_0^2} - 2\frac{a_2}{a_0}.$$

Isn't this a lovely development? I particularly like the previous identity

$$\frac{1}{r_1} + \frac{1}{r_2} + \cdots + \frac{1}{r_n} = \frac{-a_1}{a_0}.$$

To obtain it, we needed just two ideas: how coefficients depend on roots, and that reversing a polynomial inverts its roots. Each is simple to understand, but combining them we derive an unexpected result: for any polynomial, the sum of the reciprocals of the roots is a simple combination of the constant and linear coefficients. This result may seem to be as irrelevant as it is unexpected. What good does it do to know the reciprocal root sum? Do not underestimate the value of an algebraic identity. In the hands of a master like Leonhard Euler, it may prove to be just what is needed to derive astonishing new results. (See Sidebar (3.4)).

3.3   Newton's Identities

As mentioned earlier, Newton found a general scheme for expressing sums of powers of roots of a polynomial in terms of its coefficients. We have already seen some examples of such identities for a polynomial of degree n, including

$$r_1 + r_2 + \cdots + r_n = -a_{n-1}$$

and

$$r_1^2 + r_2^2 + \cdots + r_n^2 = a_{n-1}^2 - 2a_{n-2}.$$

Now we will see how these results can be extended using what are known as *Newton's identities*.

For any integer k, let s_k be the sum of the kth powers of the roots. That is,

$$s_k = r_1^k + r_2^k + \cdots + r_n^k.$$

Each s_k is a symmetric function of the r's, since reordering the r's does not change s_k. Moreover, for $k \geq 0$, each s_k is a polynomial, and is therefore expressible as a combination of the elementary symmetric functions, or equivalently, as a combination of the coefficients.

For small k, the appropriate combinations are easy to find. The derivations of $s_0 = n$ and $s_1 = -a_{n-1}$ are trivial, and $s_2 = a_{n-1}^2 - 2a_{n-2}$ is quickly established. However, for higher values of k the situation becomes more complex. Even the derivation of

$$s_3 = -a_{n-1}^3 + 3a_{n-1}a_{n-2} - 3a_{n-3}$$

requires some work. Going beyond $k = 3$ is quite a challenge.

Newton's identities provide a powerful tool for meeting this challenge. They do not give s_k directly in terms of the a's, but allow us to express s_k as a combination of preceding s's *and* the a's. Thus, s_3 is given in terms of s_1 and s_2 (as well as the a's), s_4 is given in terms of s_1, s_2, and s_3 (plus the a's), and so on. Then it is possible to relate s_k directly to the a's recursively. For example, since we know formulas for s_0, s_1, s_2, and s_3 as functions of the a's, we can substitute in the appropriate Newton's identity to obtain a corresponding formula for s_4.

In succinct form, Newton's identities can be stated as:

$$\sum_{j=0}^{n} a_{n-j} s_{k-j} = 0 \ \text{ for } k \geq n \tag{6}$$

$$\sum_{j=0}^{k} a_{n-j} s_{k-j} = (n-k)a_{n-k} \ \text{ for } k < n. \tag{7}$$

For notational simplicity, the first sum includes a_n, but we will continue to assume that $a_n = 1$.

Although compact, these equations do little to illuminate the inherent pattern of the identities. Let us look at some specific examples. Take $n = 6$ and $k = 8$. This corresponds to

Euler, Reciprocal Roots, and $\pi^2/6$

As recounted by Dunham [42, Chapter 3], the identity for the sum of reciprocal roots was used by a young Leonhard Euler to resolve a conundrum that had puzzled mathematicians for nearly fifty years. For the full story, we have to go back to 1689, and the work of Jakob Bernoulli, one of the leading mathematicians of his day. He had studied infinite sums and found the values of quite a number of them. For example, he determined that

$$\frac{1}{2} + \frac{4}{4} + \frac{9}{8} + \frac{16}{16} + \frac{25}{32} + \cdots = 6$$

and that

$$\frac{1}{2} + \frac{8}{4} + \frac{27}{8} + \frac{64}{16} + \frac{125}{32} + \cdots = 26.$$

In both sums, the denominators of the fractions are successive powers of 2. The numerators are perfect squares in the first sum, and perfect cubes in the second. Another Bernoulli result is

$$\frac{1}{3} + \frac{6}{21} + \frac{11}{147} + \frac{16}{1029} + \frac{21}{7203} + \cdots = \frac{77}{108}.$$

This time, each denominator is 3 times a power of 7, and the numerators form an arithmetic progression with difference 5.

Although Bernoulli succeeded in evaluating a great many infinite sums, there was one that evaded his grasp:

$$\frac{1}{1} + \frac{1}{4} + \frac{1}{9} + \frac{1}{16} + \frac{1}{25} + \cdots .$$

Frustrated in his attempts to find it, he issued what has become a famous challenge to the mathematical world: *If anyone finds and communicates to us that which thus far has eluded our efforts, great will be our gratitude.*

That challenge went unanswered from 1689 until 1735, when Euler stunned the world with a totally unexpected answer. The sum of the reciprocal square integers is $\pi^2/6$. This was completely unprecedented, and immediately vaulted Euler into the top ranks of mathematics. He was 28 years old.

Sidebar 3.4

the case $k \geq n$ so expand (6) to reach

$$s_8 + a_5 s_7 + a_4 s_6 + a_3 s_5 + a_2 s_4 + a_1 s_3 + a_0 s_2 = 0. \tag{8}$$

This almost looks like a polynomial, except that the s's have subscripts instead of exponents. In fact, if we replace each s_k with x^k, the left side is exactly $x^2 p(x)$. This gives an immediate verification of the identity. We know that $r_j^2 p(r_j) = 0$ for each root r_j. Therefore,

$$r_j^8 + a_5 r_j^7 + a_4 r_j^6 + a_3 r_j^5 + a_2 r_j^4 + a_1 r_j^3 + a_0 r_j^2 = 0.$$

Summing over all the roots gives (8). The same argument applies for any $k \geq n$ to establish (6).

Euler, Reciprocal Roots, and $\pi^2/6$ (cont.)

Bernoulli   Euler

What was Euler's method? He used the identity for the sum of reciprocal roots of a polynomial. However, he applied it to an infinite polynomial, that is, to a power series. He used the series

$$\frac{\sin x}{x} = 1 - \frac{x^2}{3!} + \frac{x^4}{5!} - \frac{x^6}{7!} + \cdots$$

which has roots at $x = \pm k\pi$ for nonzero integers k. But the right side has only even powers of x. Therefore, we can think of it as an infinite polynomial in $t = x^2$, and with roots given by $t = k^2\pi^2$.

Now consider the reciprocal root sum. The reciprocal roots are $1/k^2\pi^2$, and their sum is $1/\pi^2$ times the one Bernoulli wished so fervently to find. On the other hand, our identity tells us that this is $-a_1/a_0$, where the a's are the two lowest order coefficients of the polynomial. Read them off: $a_0 = 1$, $a_1 = -1/3! = -1/6$, so $-a_1/a_0 = 1/6$. In this way, Euler reached the conclusion already stated: the sum of the reciprocal square integers is $\pi^2/6$.

By today's standards, this is not a valid proof — we cannot treat an infinite power series as if it were a finite polynomial. Nevertheless, this method led Euler to a correct answer, as he himself argued with at least two other proofs later. Dozens of other proofs have been given in the intervening years. Euler's argument is breathtaking, even if it is not completely rigorous. And it illustrates my adage: never underestimate the power of an identity.

The $k < n$ case is more interesting. Again, to see the pattern it is helpful to look at specific examples. When $n = 6$ and for k decreasing from 5 to 1, expanding (7) gives

$$s_5 + a_5s_4 + a_4s_3 + a_3s_2 + a_2s_1 + a_1s_0 = a_1 \tag{9}$$

$$s_4 + a_5s_3 + a_4s_2 + a_3s_1 + a_2s_0 = 2a_2 \tag{10}$$

$$s_3 + a_5s_2 + a_4s_1 + a_3s_0 = 3a_3$$

$$s_2 + a_5s_1 + a_4s_0 = 4a_4$$

$$s_1 + a_5s_0 = 5a_5. \tag{11}$$

Here the left sides resemble polynomials (if s_k is again replaced by x^k) with a diminishing number of terms. It is neither obvious why such a pattern should hold, nor why anyone would think of it. On the other hand, there should be something familiar about the pattern above. Do you recognize it? Presently, we will see how this familiar pattern leads to a proof of (7). But first, let us examine how such an equation might have been observed in the first place.

The key idea for the $k \geq n$ case was to look at $r^{k-n} p(r)$ with r a root of p. What happens if we try this with $k < n$? For concreteness, let us take $n = 6$ and $k = 5$. Also, assume $a_0 \neq 0$, so that none of the roots of p is 0. Then, for each root r_j we know $r_j^{-1} p(r_j) = 0$, so

$$r_j^5 + a_5 r_j^4 + a_4 r_j^3 + a_3 r_j^2 + a_2 r_j + a_1 + a_0 \frac{1}{r_j} = 0.$$

As before we can sum over all the roots of p, deriving

$$s_5 + a_5 s_4 + a_4 s_3 + a_3 s_2 + a_2 s_1 + a_1 s_0 + a_0 s_{-1} = 0.$$

In the final term, s_{-1} is the sum of the reciprocals of the roots, which we found earlier to be $-a_1/a_0$ using the reverse polynomial. Therefore, substitution gives

$$s_5 + a_5 s_4 + a_4 s_3 + a_3 s_2 + a_2 s_1 + a_1 s_0 - a_1 = 0,$$

reproducing (9).

A similar approach with $k = 4$ leads to (10). Although the algebra gets progressively more involved, it is feasible to work through a few more cases so the pattern for $k < n$ can be discovered. But these examples do not point the way to a proof of the general case. Rather, it is the pattern of the equations themselves that leads to a proof.

The left sides of (9)–(11) bring to mind Horner's form and synthetic division. In fact, replacing s_k with a^k produces precisely the coefficients of the quotient $p(x)/(x - a)$, as discussed on page 9. Now when a is a root of p, say r_j, we know that $p(x)/(x - r_j)$ leaves no remainder. Thus,

$$
\begin{aligned}
\frac{p(x)}{x - r_j} = x^5 \\
+ (r_j + a_5)x^4 \\
+ (r_j^2 + a_5 r_j + a_4)x^3 \\
+ (r_j^3 + a_5 r_j^2 + a_4 r_j + a_3)x^2 \\
+ (r_j^4 + a_5 r_j^3 + a_4 r_j^2 + a_3 r_j + a_2)x \\
+ (r_j^5 + a_5 r_j^4 + a_4 r_j^3 + a_3 r_j^2 + a_2 r_j + a_1).
\end{aligned}
$$

Next, we can reintroduce the s's by summing the over all roots. The result is

$$
\begin{aligned}
\frac{p(x)}{x - r_1} + \frac{p(x)}{x - r_2} + \cdots + \frac{p(x)}{x - r_6} = 6x^5 \\
+ (s_1 + a_5 s_0)x^4 \\
+ (s_2 + a_5 s_1 + a_4 s_0)x^3 \\
+ (s_3 + a_5 s_2 + a_4 s_1 + a_3 s_0)x^2 \\
+ (s_4 + a_5 s_3 + a_4 s_2 + a_3 s_1 + a_2 s_0)x \\
+ (s_5 + a_5 s_4 + a_4 s_3 + a_3 s_2 + a_2 s_1 + a_1 s_0).
\end{aligned}
\tag{12}
$$

We can now see the connection between Newton's identities and the right-hand side of (12). Meanwhile, what can we say about the left-hand side of (12)? Simply that

$$\frac{p(x)}{x - r_1} + \frac{p(x)}{x - r_2} + \cdots + \frac{p(x)}{x - r_6} = p'(x). \tag{13}$$

This follows from the factored form of $p(x)$ as $(x - r_1)(x - r_2) \cdots (x - r_6)$ using either the product rule or logarithmic differentiation. But we know that $p'(x) = 6x^5 + 5a_5x^4 + 4a_4x^3 + 3a_3x^2 + 2a_2x + a_1$, so equating the coefficients of powers of x gives us precisely (9)–(11). This verifies Newton's identities for $k < n$.

An Alternative Approach. Newton's identities are the classical method for recursively generating the terms $s_1, s_2, s_3, \ldots$, but there is a beautiful alternative that is ideally set up for paper and pencil computation. With it you can produce the terms s_k using long division. Here is how it works. Let a polynomial $p(x)$ be given in descending form, and compute the derivative $p'(x)$. Reverse both polynomials, and then compute $\operatorname{rev} p'(x)/\operatorname{rev} p(x)$ using polynomial division, as in Sidebar I.1. The result will be the power series

$$s_0 + s_1x + s_2x^2 + \cdots .$$

Here is an example to illustrate the method. Begin with $p(x) = x^3 - 3x^2 - 4x + 12$, whose roots are 2, 3, and -2. The derivative is $p'(x) = 3x^2 - 6x - 4$. Reverse each of these, and divide. The first several steps of the division appear below.

$$
\begin{array}{r|rrrrrrr}
& 3 & + & 3x & + & 17x^2 \\
\hline
1 - 3x - 4x^2 + 12x^3 & 3 & - & 6x & - & 4x^2 \\
& 3 & - & 9x & - & 12x^2 & + & 36x^3 \\
\hline
& & & 3x & + & 8x^2 & - & 36x^3 \\
& & & 3x & - & 9x^2 & - & 12x^3 & + & 36x^4 \\
\hline
& & & & & 17x^2 & - & 24x^3 & - & 36x^4.
\end{array}
$$

Notice that everything appears in ascending order. This makes sense because the quotient, $q(x) = \operatorname{rev} p'(x)/\operatorname{rev} p(x)$ will be a rational function with an infinite power series. We obviously cannot generate the terms working down from the highest power of x, so we work from the lowest power up, instead. The illustration only shows the first three terms of q, but it is clear that the process can be continued to generate as many terms as we wish.

And the point of the computation is that we can now read off $s_0 = 3$, $s_1 = 3$, $s_2 = 17$, and so on, from the expansion of $q(x)$. For this example we can verify that the results are correct because we know the roots of p are -2, 2, and 3. First compute the sum of zeroth powers of the roots. That is, $s_0 = (-2)^0 + 2^0 + 3^0 = 3$, which is the constant coefficient of q. Next, $s_1 = (-2)^1 + 2^1 + 3^1 = 3$, and that agrees with the linear coefficient of q. Finally, $s_2 = (-2)^2 + 2^2 + 3^2 = 17$, agreeing with the quadratic coefficient of q.

This process for generating the terms s_k is easy to understand and easy to carry out. But why does it work? What does division of the reverse polynomial and derivative have to do with sums of roots of p? For the answer, we go back to the earlier proof of Newton's identities. From (13), we see that

$$\frac{p'(x)}{p(x)} = \frac{1}{x - r_1} + \frac{1}{x - r_2} + \cdots + \frac{1}{x - r_n} \tag{14}$$

for any polynomial p of degree n with roots $r_1, r_2, \ldots, r_n$. But we are interested in the ratio of the reversed polynomials. Recalling that $\operatorname{rev} p(x) = x^n p(1/x)$, the ratio is

$$\frac{\operatorname{rev} p'(x)}{\operatorname{rev} p(x)} = \frac{x^{n-1} p'(1/x)}{x^n p(1/x)} = \frac{1}{x} \cdot \frac{p'(1/x)}{p(1/x)}.$$

Thus, replacing x by $1/x$ in (14), we obtain

$$\frac{\operatorname{rev} p'(x)}{\operatorname{rev} p(x)} = \frac{1}{x} \cdot \left(\frac{1}{1/x - r_1} + \frac{1}{1/x - r_2} + \cdots + \frac{1}{1/x - r_n} \right)$$

$$= \frac{1}{1 - r_1 x} + \frac{1}{1 - r_2 x} + \cdots + \frac{1}{1 - r_n x}.$$

Next, we can express each fraction on the right as a geometric series, because

$$\frac{1}{1 - r_k x} = 1 + r_k x + r_k^2 x^2 + r_k^3 x^3 + \cdots.$$

Combining these geometric series gives us

$$n + (r_1 + r_2 + \cdots + r_n)x + (r_1^2 + r_2^2 + \cdots + r_n^2)x^2 + \cdots = s_0 + s_1 x + s_2 x^2 + \cdots.$$

This shows that

$$\frac{\operatorname{rev} p'(x)}{\operatorname{rev} p(x)} = s_0 + s_1 x + s_2 x^2 + \cdots,$$

justifying the long division approach for generating the s_k.

An Application.   The following problem appeared in the April 2006 issue of *Math Horizons*:

Prove that $\sqrt[7]{13 + \sqrt{41}} + \sqrt[7]{13 - \sqrt{41}}$ is not a rational number.

Here is a solution using root power sums. Let $\alpha = \sqrt[7]{13 + \sqrt{41}}$ and $\beta = \sqrt[7]{13 - \sqrt{41}}$, so that $\alpha\beta = \sqrt[7]{13^2 - 41} = 2$. We wish to show that $\alpha + \beta$ is not rational. Thinking about the sum and product of α and β brings to mind an earlier observation about quadratic polynomials (see page 50): α and β are the roots of $p(x) = x^2 - (\alpha + \beta)x + \alpha\beta$. If we let $b = \alpha + \beta$, $p(x) = x^2 - bx + 2$. And now, since α and β are the roots of p, we can use the division method to find $\alpha^7 + \beta^7 = s_7$ in terms of the coefficients of p. Since, by inspection, $\alpha^7 + \beta^7 = 26$, this leads to an equation involving b. All we need is the formula for s_7.

Let us apply the division method. Reversing $p(x) = x^2 - bx + 2$ and $p'(x) = 2x - b$, we must compute $(2 - bx)/(1 - bx + 2x^2)$. The first few steps look like this:

$$
\begin{array}{r|rrrrrr}
 & 2 & + & bx & + & (b^2 - 4)x^2 & \\
\hline
1 - bx + 2x^2 & 2 & - & bx & & & \\
 & 2 & - & 2bx & + & 4x^2 & \\
\hline
 & & & bx & - & 4x^2 & \\
 & & & bx & - & b^2 x^2 & + & 2bx^3 \\
\hline
 & & & & & (b^2 - 4)x^2 & - & 2bx^3.
\end{array}
$$

Reading off the coefficients, we observe that

$$s_0 = 2$$
$$s_1 = b$$
$$s_2 = b^2 - 4.$$

Continuing the long division, we eventually come to

$$s_7 = b^7 - 14b^5 + 56b^3 - 56b.$$

But we already observed that $s_7 = 26$. Thus, we have derived the equation

$$b^7 - 14b^5 + 56b^3 - 56b - 26 = 0.$$

This shows that if b is rational, it must be a rational root of $q(x) = x^7 - 14x^5 + 56x^3 - 56x - 26$. Since q is monic with integer coefficients, the only possible rational roots are integer divisors of 26. (This is the Rational Roots theorem, which will be discussed in Section 4.1.) Accordingly, if b is rational, it must be in the set $A = \{\pm 1, \pm 2, \pm 13, \pm 26\}$.

We know that $\alpha^7 = 13 + \sqrt{41}$, which is between 19 and 20. This shows that $1 < \alpha < 2$. A similar computation shows that $1 < \beta < 2$. Since $b = \alpha + \beta$, we have $2 < b < 4$. This shows that b is not in the set A, and so cannot be rational.

Although our algorithm for computing s_k values is just the right tool for the preceding problem, I am not suggesting that this establishes the importance of the s_k or Newton's identities, or that it justifies their inclusion here. After all, one might easily question whether the application itself is interesting or important. In fact, this application is included purely due to serendipity. The problem showed up in my email one day, about the time I was writing this chapter. Because I was thinking about Newton's identities, the s_k sequence came naturally to hand in trying to solve the problem. So, if this application illustrates anything, it is that root identities like Newton's are handy to have in your problem solving toolbox: you never know when they might prove useful.

Besides that, the identities discussed in this section are intrinsically interesting. I like the contrast between the difficulty of finding roots of polynomials and the ease with which we can obtain identities involving combinations of them. Another part of the appeal is the appearance of reverse polynomials and reciprocal roots; likewise the reappearance of synthetic division and Horner evaluation.

The symmetry of functions that are invariant under permutation of their variables is a key feature of the material we have considered in this chapter. This idea will occur also in the next chapter when we consider the algebraic solution of cubic, quartic, and higher degree equations.

3.4 History, References, and Additional Reading

An excellent resource both for history and deeper mathematical development is Edwards' book on Galois theory [47]. Galois theory is the modern framework for understanding how and when roots of polynomials can be constructed algebraically. This topic will be mentioned again in the next chapter, as well. The elementary symmetric functions that we have considered in this chapter are important in the development of Galois theory. In fact, Edwards emphasizes what he calls the fundamental theorem on symmetric polynomials:

Any symmetric polynomial in the roots of an equation can be expressed in terms of the coefficients of that equation. To Edwards, this theorem is "the foundation stone of Galois theory." For this reason, he begins his book with a careful discussion of the theorem, and presents many historical details about its development. I recommend the first thirteen sections of Edwards' book for these topics. Edwards was my principal source for Sidebar (3.2).

Edwards gives a detailed proof of the fundamental theorem, remarking that it can be used as an algorithm for actually finding the representation of a given symmetric polynomial in terms of the coefficients. He cautions that this is not particularly practical. Another source for the fundamental theorem is van der Waerden [161, pp. 99–101]. This treatment also presents a proof that can be used as an algorithm, and the steps are explicitly given, but I suspect that this algorithm suffers the same practical limitations for high degree polynomials as Edwards'. However, even if that is true, van der Waerden's algorithm is a helpful guide for experimental investigation using a computer algebra system, as discussed in Sidebar (3.3).

Both Edwards and van der Waerden are also good sources for additional reading on Newton's identities and the discriminant of a polynomial. Edwards gives details of Newton's discovery and publication of these identities, including nineteen identities for symmetric polynomials for three roots. According to Edwards, Newton's first published statement of his identities was in the *Arithmetica Universalis* of 1707, although historical investigations suggest that Newton was aware of the full result during the mid-1660's. Also, Newton was apparently unaware of similar or equivalent results of Girard dating to 1629.

A variety of proofs have been published for Newton's identities, the shortest of which is surely Zeilberger's paper [172]. My paper [79] offers a proof using matrix methods, and includes references to several other proofs. The idea of relating the series $s_0 + s_1 x + s_2 x^2 + \cdots$ to the ratio of the reversed polynomial and the reversed derivative is easy to find in the literature ([14, 124]). I first learned of this in [30], where the idea of using long division to generate the s_k's is explicitly given.

On the subject of symmetry, Stewart's book [153] is an excellent source for further study. It provides a detailed look at the evolution of our modern concept of symmetry, including a long historical discourse on solving polynomial equations. Another source is Lederman and Hill's [108], which discusses symmetry in the context of science and especially physics.

4

Solving Polynomial Equations

Finding the roots of polynomials is a problem with a long history in mathematics, and one that has led to a tremendous volume of mathematical knowledge. In elementary mathematics we first encounter quadratic polynomials, and learn to find roots using factoring or the quadratic formula. The outcome depends on what types of numbers you are willing to accept as answers. If you accept complex numbers, then the quadratic formula gives a complete solution. Every quadratic has two roots (sometimes equal), and complete factorization is always possible. In contrast, if you are only interested in real solutions, some quadratic equations are unsolvable. Here again the quadratic formula gives us a complete answer. It tells us whether roots exist, and finds them when they do.

It is natural to turn next to cubic equations. We can find solutions for quadratic equations, like $x^2 - 3x + 5 = 0$. The cubic equation $x^3 + 7x^2 - 2x + 3 = 0$ doesn't look much more complicated. Can we solve it? Is there some way to find the solutions as combinations of cube roots and square roots, perhaps, extending what works for quadratics? And beyond cubics, what about quartics, quintics, and higher degree polynomials?

The attempt to solve cubics dates back at least to the tenth century efforts of Arab mathematicians. The earliest work had a strong geometric component, both in how the equations were understood and in methods of solution. Omar Khayyam discovered how to find roots of cubics as intersections of parabolas [3]. An algebraic solution for one form of cubic was developed in the beginning of the sixteenth century by Cardano, among others. Complete solutions for general cubic and quartic equations followed soon after. The cubic and quartic solutions have a similar flavor. To solve a cubic, you must first solve a related quadratic, and the solutions of the cubic involve square and cube roots. Similarly, the solution of a quartic depends on solving a related cubic.

We will see these solutions in this chapter. By today's standards, the algebra they involve does not seem very forbidding. The solution to cubic equations can be followed by students at the level of precalculus. Of course, following the logic of an algebraic argument is far easier than inventing the argument in the first place. And the first discovered solutions to cubics and quartics involve clever algebraic tricks. Moreover, they came before the development of modern algebraic notation, which makes the solutions easier to follow. Still, it is tempting to wonder why the solutions took so long to be found, and to imagine that similarly simple methods for solving higher equations might yet await discovery.

Interestingly, the pattern of results for polynomials up to the quartic do not extend to higher degrees. More subtle patterns lurk behind the scenes that account for both the solutions of the low degree cases and the obstructions to finding similar solutions for higher degree. A key to understanding this deeper structure involves the interchangeability of the roots of a given polynomial. As we will see, this idea arises quite naturally at a very elementary level. We have already seen it emerge in the previous chapter's study of the relationships between roots and coefficients. The significance of the idea of interchangeability goes far beyond these simple beginnings, however. It ultimately unlocks the deep mysteries of polynomial equations in a mathematically breathtaking development. The end of the chapter will offer a glimpse of this subject.

The quest to solve polynomial equations has played a central role in the evolution of modern mathematics. Along the way, mathematicians have encountered important methodological and philosophical questions. What does it mean to *find* a root? If we can approximate a root to any specified accuracy, can we then claim to have found it? Can we even claim to know that the root exists? These questions go beyond solving equations to the very nature of numbers themselves, and historically led to ever richer number systems, including negatives, irrationals, and imaginary numbers.

In broad outlines, here is what we know today. Permitting complex numbers to be used, both for coefficients and roots, every polynomial can be expressed as a product of linear factors. Thus, for any polynomial of degree n, there exist n complex roots. When n is 4 or less, exact solutions can be found using known methods, which are analogous to the quadratic formula and can be applied to any cubic or any quartic. For polynomials of degree 5 or higher, it is known that the roots do not always exist in an exact form, or at least, not as an algebraic expression involving arithmetic operations and radicals (i.e., square roots, cube roots, fourth roots, and so on). This shows that a general method for solving quintic and higher degree equations cannot exist.

These conclusions are valid when working with the complex numbers. But polynomials often arise where coefficients are known to be real, or rational, or integers, or where we are interested only in real, rational, or integer roots. This leads to quite a few different possible problems and conclusions. For example, if we have integer coefficients and desire only rational roots, an exact solution method for arbitrary degree is known.

In this chapter we will look at a variety of ideas connected with solving polynomial equations, including:

- Existence questions for rational, real, and complex solutions,

- Algebraic solutions of cubic and quartic equations,

- Lagrange's analysis and permutations of roots,

- Insolubility of the general quintic.

4.1   Existence Questions

Before talking about how to solve a polynomial equation, we should clarify what it means for a solution to exist. If it is possible to display an integer or rational number that satisfies a given equation, then the question of existence of course evaporates. But what about equations with irrational or complex solutions?

Today, with the complex numbers a familiar and well understood part of mathematics, we have an advantage over mathematicians of earlier eras. Our current ideas about number systems, indeed about what numbers *are*, evolved slowly over centuries. At the time of Cardano, negative numbers were sometimes used, but were not universally accepted as legitimate numbers. Complex numbers were contemplated only by the most advanced thinkers, and even they did not know what to make of the idea. Little wonder if there was some ambiguity about what it means for an equation to have a solution.

To make this more concrete, let us consider a few examples. According to the modern viewpoint, the equation $x^2 = 1$ has two solutions. A mathematician who denies the existence of negative numbers would say it has a single solution. Cardano would have identified two solutions, calling one *false*, meaning not really a legitimate number. The equation $x^2 + 1 = 0$ would be considered to have no solutions whatever.

Similar issues arise with irrational numbers. As an example, consider the equation $3x^5 - 15x + 5 = 0$. It has no rational roots (which can be verified by methods to follow), but by trial and error (as well as more sophisticated methods), we can find approximate solutions. To illustrate one approach, let $p(x) = 3x^5 - 15x + 5$, so $p(0) = 5$ and $p(1) = -7$. This suggests that a root should occur somewhere between 0 and 1. We might guess that it falls at $x = .5$. This guess is incorrect, but with a true solution somewhere between 0 and 1, $x = .5$ differs from a correct answer by at most .5.

Next compute $p(.1)$, $p(.2)$, $p(.3)$, and so on, finding $p(.3) = .50729$ and $p(.4) = -.96928$. This tells us the root is between .3 and .4, and if we adopt .35 as an estimate, we know we will be off by at most .05. Continuing in this way, we can get as close to the root as we wish.

But does this establish the existence of an exact solution? If there is no specific number we can identify that is a solution, if we can only point to better and better approximations, how do we know that there *is* a solution?

The modern viewpoint rests on a sophisticated understanding of both the algebraic and geometric properties of our number systems. We have a concept of continuity, according to which the real numbers form a continuum with no holes or gaps. This idea can be formulated rigorously and, as long as you are willing to accept the assumptions inherent in the formulation, results such as the intermediate value theorem give precise conditions for the existence of solutions to equations.

Historically, solving polynomial equations meant finding exact algebraic expressions for the roots. In the next section the original methods for solving cubic and quartic equations in this sense will be presented. Generalizing them to higher degrees involves the concept of *solvability by radicals*. That means solutions that are expressible in terms of radicals (square roots, cube roots, etc.) and arithmetic operations. The methods for cubics and quartics show they are all solvable by radicals. For higher degree polynomials, some equations are solvable by radicals, and some are not.

Is a solution by radicals preferable to one by successive approximation? Aesthetically, an exact algebraic representation of the solution is highly appealing. But the more aesthetic approach is not always the more practical. Working on a computer or calculator with limited precision, successive approximation is typically more efficient and accurate than direct implementation of solutions by radicals, at least for degree 3 or 4. Even if there is an exact algebraic expression for a root, it most likely will involve radicals that are themselves only approximately computable. And on the aesthetic side solution by radicals is not the only option, for we saw exact solutions to cubic equations elegantly expressed in terms of curly roots (page 26).

As these considerations show, solvability of polynomial equations is a many-faceted subject. We can approach it in a variety of ways, conceptually, procedurally, and philosophically. With that in mind, let us look at some important solvability results.

Rational roots.   Suppose that a polynomial has coefficients that are rational numbers. What can we deduce about the roots? First, any such equation can be transformed into one with integer coefficients. Expressing each of the original coefficients as a fraction, we can find a common multiple A of all the denominators. Multiplying the entire equation by A then eliminates all the denominators, leaving a polynomial with integer coefficients.

Fair enough — let us restrict our attention to polynomials with integer coefficients. Solutions may be rational or irrational. But if they are rational, they have to take a special form, as specified in the following result.

The Rational Roots Theorem.   *Let $p(x) = a_n x^n + a_{n-1} x^{n-1} + \cdots + a_0$ have integer coefficients, and for integers r and s let r/s be a rational root in lowest terms. Then r is a divisor of a_0 and s is a divisor of a_n.*

Proof.   Since r/s is a root, $p(r/s) = 0$, hence $s^n p(r/s) = 0$. This gives us the equation

$$a_n r^n + a_{n-1} r^{n-1} s + a_{n-2} r^{n-2} s^2 + \cdots + a_1 r s^{n-1} + a_0 s^n = 0,$$

which can be written in the form

$$a_n r^n + a_{n-1} r^{n-1} s + a_{n-2} r^{n-2} s^2 + \cdots + a_1 r s^{n-1} = -a_0 s^n.$$

Because r is a divisor of the left side of this equation, it is a divisor of $a_0 s^n$. But r/s is in lowest terms means that r and s have no common divisors. Thus r is a divisor of a_0. A similar argument shows that s is a divisor of a_n. ∎

Ideally, we would like our equations to have rational solutions. The preceding theorem provides a tool for investigating whether or not they do. Rational roots, if they exist, have to lie among the finitely many fractions whose numerators divide into a_0 and whose denominators divide into a_n. If we compute $p(x)$ for each of these fractions, we will find all the rational roots, or demonstrate that there are none.

As an example, consider $x^3 + 6x - 20 = 0$. The divisors of a_3 are just 1 and -1. The divisors of a_0 are $1, 2, 4, 5, 10, 20$, and their negatives. Therefore, if the equation has rational roots, they must lie among the numbers $\pm 1, \pm 2, \pm 4, \pm 5, \pm 10$, and ± 20. Direct calculation reveals that 2 is a rational root, and it is the only rational root.

This can be verified from another direction. Knowing that $x = 2$ is a root implies that $x - 2$ is a factor of the polynomial, and it is a short step to $x^3 + 6x - 20 = (x - 2)(x^2 + 2x + 10)$. The quadratic formula shows that the roots of the quadratic factor are $-1 \pm 3i$. This both verifies that 2 is the only rational root and illustrates the important idea of using a known root to reduce the degree of an equation.

The example also illustrates a corollary of the Rational Roots Theorem: if a *monic* polynomial has integer coefficients, any rational roots are actually integers. As the example shows, when the leading coefficient of a polynomial is 1, the denominator of any rational root must be ± 1. In more advanced treatments of polynomials, roots of monic polynomials are referred to as *algebraic integers*, and play a special role in the development of the theory.

Real roots. As many polynomials with integer coefficients have irrational roots, it is natural next to extend our focus to real numbers. Whereas the rational case is characterized by discrete methods (integer factorization, enumerating a finite set of possible solutions), the real case has a decidedly continuous flavor. For example, using the continuity of the real line and of polynomial functions, we can invoke the intermediate value theorem to show that if $p(x)$ has opposite signs at a and at b then it has a root somewhere in between. This is an existential result. It does not tell us how to find a root, only that one must exist. But that is an important consideration in locating roots through successive approximation.

It is also customary to use methods from calculus in the real case. A function that is increasing in an interval, for example, can have at most one root in that interval. Here, an *increasing function* $f(x)$ obeys the maxim *the larger the x, the larger the $f(x)$*. If $f(x) = 0$ for a specific x, then it must be greater than 0 for all larger values of x, so there can be at most one root.

A familiar result from calculus tells us that a function is increasing if it has a positive derivative. Let us apply this idea to the earlier example, with $f(x) = x^3 + 6x - 20$. The derivative, $f'(x) = 3x^2 + 6$, is positive for all real x. Therefore, $f(x)$ is increasing over the entire real line, and can have at most one real root. This shows that $x = 2$ is not just the only rational root, but also the only *real* root.

Similar analyses lead to a number of conclusions for polynomials with real coefficients:

- An odd degree polynomial always has at least one real root.

- An even degree polynomial has an even number of real roots (where a double root is counted twice).

- For odd n, the equation $x^n - a_0 = 0$ always has exactly one real root.

- For even n, the equation $x^n - a_0 = 0$ has two real roots if a_0 is positive, the unique root $x = 0$ if $a_0 = 0$, and no real roots if a_0 is negative.

The last two of these give us familiar properties about nth roots.

Complex roots. The complex numbers are the ultimate setting for polynomial equations in a particular sense. Observe the progression of number systems. Simple (linear) equations with natural number coefficients can have solutions that are negative, or noninteger rationals. So we extend the number system to include negatives and fractions, arriving at the rational numbers. Polynomial equations with rational coefficients can have irrational solutions, so we extend the number system again, this time to the reals. But there are equations with real coefficients that do not have real solutions. This leads us to the complex numbers. Now the process stops, because every polynomial with complex coefficients has roots that are complex numbers. At least in terms of formulating and solving polynomial equations, the complex numbers form a closed system. Technically, the complex numbers are said to be *algebraically closed*.

This result is known as the *Fundamental Theorem of Algebra*. Here is a formal statement.

The Fundamental Theorem of Algebra.   *A non-constant polynomial with complex co-efficients has a root in the complex numbers.*

The existence of a root also implies a factorization. If a polynomial $p(x)$ has a complex root r_1, then we can factor it in the form $p(x) = (x - r_1)q(x)$. Since $q(x)$ will also have complex coefficients, it too must have a root (unless it is a constant), and it will have degree one less than p. By applying the Fundamental Theorem in a chain of factorizations, we will eventually reduce $p(x)$ to a factored form $A(x - r_1)(x - r_2) \cdots (x - r_n)$. Thus, the Fundamental Theorem shows that every polynomial equation with complex coefficients can be completely factored, and so has all of its roots among the complex numbers.

As in the real case, this is an existence result. And like the real case, it reflects not only the algebraic properties of the complex numbers, but also the important idea of continuity. This is a topological or analytic aspect of the complex numbers. Many proofs of the Fundamental Theorem are known, and they all involve analysis or topology in some way.

Earlier we described the complex numbers as the ultimate number system for polynomial equations. It really would be more accurate to say *an* ultimate system. In other contexts polynomial equations arise with coefficients that are outside the integer-rational-real-complex progression. For example, polynomial equations with matrix coefficients arise naturally in some areas of mathematics. Another example is provided by modular arithmetic. It makes perfect sense to discuss the solvability of polynomial equations whose coefficients are integers modulo five, and that leads to a completely different ultimate number system.

Even so, it remains a valid observation that in familiar number systems, namely those encompassing the integers, the complex number system is the one that has all the answers. It provides the proper perspective for understanding polynomials, even in cases that seem to involve only real numbers. For example, the algebraic solution of cubic equations requires complex numbers, even when all the coefficients and roots are real. Indeed, it has been argued that solving cubic equations provided a primary motivation for the historical development of complex numbers [99].

We close this section with an illustration of complex numbers intruding on a real issue. It is a theorem about real factorizations of real polynomials, but the proof uses complex numbers.

The Fundamental Theorem of Algebra, Real Case.   *Every polynomial with real coefficients can be expressed as a product of linear and quadratic factors with real coefficients.*

This is a corollary of the Fundamental Theorem of Algebra, and has a simple proof using complex conjugation. (See Appendix A at the website for this book [87] for an explanation of conjugation and other aspects of complex numbers.) In particular, if $p(x)$ has real coefficients, then for any complex number z, $p(\bar{z}) = \overline{p(z)}$. This implies that r is a root of p if and only if $\bar{r}$ is also a root, because

$$p(\bar{r}) = \overline{p(r)} = \bar{0} = 0.$$

Over the complex numbers, a polynomial with real coefficients factors as $A(x - r_1) \cdot (x - r_2) \cdots (x - r_n)$ with roots r_j that are complex numbers. Some of these may be real. For those that are not, each factor $(x - r)$ combines with a corresponding factor $(x - \bar{r})$ to form a quadratic factor with real coefficients. This establishes the real case of the Fundamental Theorem of Algebra.

4.2 Historic Solution of Cubics and Quartics

In mathematics, the first solution of a problem is often far from the best solution. Later developments can lead to solutions that are shorter and more transparent than the one first discovered. However, this is not the case for solving cubic equations (in terms of radicals). The algebraic solution first published by Cardano in 1545 is as simple as any of the many variations and alternatives that have appeared since. An example will illustrate how this solution works. We will consider an example that Cardano used.

In $x^3 = 20 - 6x$, substitute $u + v$ for x to obtain

$$(u + v)^3 = 20 - 6(u + v). \tag{1}$$

The cube on the left is $u^3 + 3u^2v + 3uv^2 + v^3$. When we group the u^3 and v^3 terms together and factor $3uv$ from the remaining terms, we have

$$u^3 + v^3 + 3uv(u + v) = 20 - 6(u + v).$$

Now we separate this one equation into two:

$$u^3 + v^3 = 20$$
$$3uv = -6.$$

Clearly, if we can find numbers u and v that satisfy these two equations, then $u + v$ will satisfy (1), and hence the original equation.

At first glance, this appears similar to our earlier attempts to solve a cubic using elementary symmetric functions (see page 50). But this time the system can be solved. In fact, a solution follows from our understanding of the elementary symmetric functions for a quadratic. To make this clearer, divide the second equation by 3, and then cube both sides. The system then takes the form

$$u^3 + v^3 = 20$$
$$u^3v^3 = -8.$$

Thinking of u^3 and v^3 as the unknowns, the system specifies their sum and product — exactly the requirements for solving a quadratic equation as discussed on page 50. So u^3 and v^3 must be the solutions of the quadratic equation

$$t^2 - 20t - 8 = 0.$$

Solving, we find that u^3 and v^3 must be $10 \pm 6\sqrt{3}$. Extracting cube roots gives us values for u and v, and leads to

$$x = \sqrt[3]{10 + 6\sqrt{3}} + \sqrt[3]{10 - 6\sqrt{3}} \tag{2}$$

as a solution to the original cubic.

In principle, once a root r is known, the factor $x - r$ can be divided out of the original cubic. The remaining roots can then be found using the quadratic formula. This is not a very practical way to proceed in general, and is quite unwieldy in the example at hand. An alternative is to recognize that extracting a cube root is itself the solving of a polynomial

Sidebar 4.1

The History of the Cubic

The solution of the cubic equation is accompanied by an intriguing human interest story, centered on a dispute between Cardano and Tartaglia. Scipione del Ferro made the first progress toward a general algebraic method for cubics sometime between 1500 and 1515. His method applied to equations of the form $x^3 + cx = d$, with both c and d assumed to be positive. Today we consider a single standard cubic $x^3 + cx + d = 0$, but in del Ferro's time negative numbers were not accepted and the terms of the equation had to be positioned so as to avoid negative coefficients. To del Ferro, $x^3 + 3x = 5$ and $x^3 + 5 = 3x$ were distinct problems, and called for distinct methods of solution.

Although del Ferro's discovery was a major breakthrough, he did not make it known within the mathematical community. Rather, he kept his method secret so that he would be able to solve equations that stumped mathematical rivals. But before his death in 1526 he passed his method on to two colleagues, Antonio Maria Fiore and Annibale della Nave. In 1535, Fiore challenged Tartaglia to a problem solving contest. Inspired by the problems in the contest, Tartaglia discovered his own method for solving cubic equations. Following del Ferro's example, he kept his method secret.

Now Cardano enters the picture. He was working on a book at the time, and wanted to include Tartaglia's method. Tartaglia eventually agreed to reveal his method to Cardano, but wishing to publish the method himself, Tartaglia made Cardano promise not to reveal it. This was in 1539. True to his word, Cardano left Tartaglia's method out of his book.

But over the next several years, he worked on the cubic himself, working out methods for all the various cases (what we would regard as possible combinations of signs of coefficients). He also learned of del Ferro's original method and obtained the

equation, and so leads potentially to three solutions. So once we know a value for u^3 we should be able to obtain three values for u, corresponding to three roots of the original cubic equation. We will follow up on this idea below.

Curiously, the solution (2) is a complicated way of writing $x = 2$, though there is no simple algebraic method to verify that fact. The simplest verification is to show directly that 2 is a root of the original equation, and, in fact, the only real root. Here we are observing the tip of an iceberg. There is much more to say about identities involving radical expressions and how they relate to roots of polynomials, but it would be too great a digression to explore this topic now. Suffice it to say that it is sometimes difficult to recognize when an algebraic expression involving radicals has a simpler equivalent form.

Does this single example convince you that Cardano's method will solve *any* cubic equation? One issue is the special form of the example, with no x^2 term. But we saw in Chapter 3 that any cubic can be brought to this reduced form by making a substitution for the variable (see page 51). Another objection could arise from the fact that u^3 and v^3 are supposed to be obtained as roots of a quadratic. What if the quadratic does not have any real roots? Then we can express u^3 and v^3 as complex numbers, but how do we find u and v?

The History of the Cubic (cont.)

Cardano

Tartaglia

details from della Nave. Since del Ferro's discovery preceded Tartaglia's, and since Cardano had extended the method to many new cases, Cardano considered himself no longer bound by his vow to keep Tartaglia's method secret. In 1545 he published the *Ars Magna* (or *The Great Art*), which included a comprehensive account of methods for all the possible cubics, as well as a method for the quartic discovered by his student, Lodovico Ferrari.

Tartaglia was outraged, and thought Cardano had betrayed his trust. To add insult to injury, in a later mathematical contest he fell to defeat at the hands of Ferrari, Cardano's protege. Intrigue, betrayal, triumph, defeat — did you ever imagine that the history of algebra could be as dramatic as a soap opera?

Sidebar 4.1 (cont.)

Algebra has developed far beyond what was known to Cardano and his contemporaries. Today we have better notation and symbolic methods, and can draw on the full power of the complex number system. And in large measure, these tools were developed for the investigation of polynomial equations.

Our modern vantage point augments the algebraic derivation of Cardano's solution with a complete understanding of the solution of the general cubic equation. We now know that a single valid choice for u and v can be used to completely factor the original cubic; that this will involve cube roots of complex numbers exactly when the cubic has only real roots; and how to construct those complex cube roots. These points will be considered in detail before we proceed to quartic equations.

As a first step of the analysis of the cubic, let us retrace Cardano's method for the general equation

$$x^3 = bx + c.$$

We introduce the substitution $x = u + v$ and derive the equations

$$u^3 + v^3 = c$$
$$uv = b/3, \tag{3}$$

Cubing the second equation produces

$$u^3 + v^3 = c$$
$$u^3 v^3 = b^3/27,$$

a system of equations in u^3 and v^3.

If we let $s = u^3$ and $t = v^3$, then the system becomes

$$s + t = c$$
$$st = b^3/27.$$

Now it is apparent that s and t are roots of a quadratic. We extract cube roots of s and t to determine u and v, and hence a solution $u + v$ to the cubic. However, there are actually three complex cube roots of s. Let u_0 be a particular cube root of s. Then the other cube roots can be expressed in terms of u_0 and $\omega = (-1 + i\sqrt{3})/2$, a primitive cube root of unity. There are exactly three complex numbers with cube equal to s, namely u_0, ωu_0, and $\omega^2 u_0$.

We do not have equal freedom to choose v among the cube roots of t. Referring to (3), once u has been selected, we must take $v = b/(3u)$, so the system of equations leads us to exactly three choices for the pair (u, v). That is, Cardano's method gives us three expressions for a root of the cubic equation: $u_0 + b/(3u_0)$, $\omega u_0 + b/(3\omega u_0)$, and $\omega^2 u_0 + b/(3\omega^2 u_0)$. If we let $u = u_0$ and $v = b/(3u_0)$, then since $\omega^3 = 1$, our three expressions for roots of the cubic become $u + v$, $\omega u + \omega^2 v$, and $\omega^2 u + \omega v$.

There is no reason to assume that the three expressions are distinct, and in fact they need not be. But we can show that they give a complete factorization for the original cubic. The elementary symmetric functions again are useful. We consider the product

$$q(x) = (x - u - v)(x - \omega u - \omega^2 v)(x - \omega^2 u - \omega v). \tag{4}$$

This can be put in descending form with coefficients given by the elementary symmetric functions of the three roots:

$$a_0 = -(u + v)(\omega u + \omega^2 v)(\omega^2 u + \omega v)$$
$$a_1 = (u + v)(\omega u + \omega^2 v) + (u + v)(\omega^2 u + \omega v) + (\omega u + \omega^2 v)(\omega^2 u + \omega v)$$
$$a_2 = -(u + v) - (\omega u + \omega^2 v) - (\omega^2 u + \omega v).$$

To simplify these algebraically, it helps to observe that $1 + \omega + \omega^2 = 0$, which follows from $0 = (\omega^3 - 1) = (\omega - 1)(\omega^2 + \omega + 1)$. The alternative form $\omega + \omega^2 = -1$ is also handy. Using these identities, the system simplifies to

$$a_0 = -u^3 - v^3$$
$$a_1 = -3(uv)$$
$$a_2 = 0.$$

Since we know that $u^3 + v^3 = c$ and $3uv = b$, it follows that $a_0 = -c, a_1 = -b$, and the descending form of $q(x)$ is thus $x^3 - bx - c$. This demonstrates that the three roots of $q(x)$ are all the roots of the original equation $x^3 = bx + c$.

It is important to stress here the dependence of the factorization just derived on an arbitrary choice of three possible values of u. For each choice of u_0, we get three roots (and a factorization) of $x^3 - bx - c$. This might seem to suggest a total of nine possible roots, but we know that there must be exactly three. Therefore, each choice of u_0 leads to the same three roots, appearing in different orders. Earlier, we saw that the coefficients of a polynomial depend symmetrically on the roots, since permuting the roots has no effect on the coefficients. But now we see how the idea of permuting the roots arises naturally from an ambiguity in defining a complex cube root. This same phenomenon occurs any time a root of a polynomial is expressed in terms of radicals. Each radical can be interpreted as one of n different complex nth roots, any choice of which leads to a permutation of the roots produced by any other choice.

Expressing the three roots as in (4) is of interest for another reason. From it we can deduce that all three roots of the cubic are real precisely when s and t are complex (that is, are not real). Because s and t arise as roots of the quadratic equation

$$x^2 - cx + b^3/27 = 0,$$

the discriminant $D = c^2 - 4b^3/27$ tells us whether or not they are real.

If $D \leq 0$, then s and t are complex conjugates (and possibly real and equal). In this case, u and v are also complex conjugates. To see this, note that $v^3 = t = \bar{s} = \overline{u^3} = \bar{u}^3$. Thus, v and $\bar{u}$ are both complex cube roots of t, so are equal or differ by a factor of ω or ω^2. That is, $v = \bar{u}$ or $v = \omega\bar{u}$ or $v = \omega^2\bar{u}$. But we also know that uv is real. Thus, $v = \bar{u}$, showing that one of the roots, $u + v$, is real. Next, observe that ω and ω^2 are complex conjugates. Thus each of the remaining roots, $\omega u + \omega^2 v$ and $\omega^2 u + \omega v$, is again the sum of a complex number and its conjugate, and so real. Therefore, when $D \leq 0$ all three roots are real.

On the other hand, suppose $D > 0$. Then s and t are real and unequal. Choosing a real value for u, and thus ensuring that v is also real, we obtain one real root, $r_1 = u + v$. A second root, $r_2 = \omega^2 u + \omega v$, can be rewritten

$$\omega^2 u + \omega u - \omega u + \omega v = -u + \omega(v - u).$$

If this is real, then $v = u$ so $s = u^3 = v^3 = t$, contrary to assumption. Thus r_2 is not real, and by a similar argument, neither is the final root, $r_3 = \omega u + \omega^2 v$.

Putting all this together, we see that a cubic has three real solutions if and only if $D \leq 0$, and three distinct real solutions when $D < 0$. This latter case occurs precisely when s and t fail to be real, and then we can reach the real solutions only by operating with complex numbers.

It is natural to ask whether this is an artifact of Cardano's method, or an intrinsic property of cubic equations. Could there be an alternative method for solving cubics that avoids complex numbers, at least when the roots are all real? Motivated partly by this question, we will look at alternate solutions of the cubic in the next section.

Now let us turn to the quartic. As mentioned in Sidebar 4.1, Cardano's student Ferrari is credited with first discovering a method for solving quartic equations. In modern form, Ferrari's method proceeds as follows.

Begin with a reduced equation (no x^3 term) written in the form

$$x^4 = ax^2 + bx + c.$$

Now introduce a new variable y and add $2x^2y + y^2$ to both sides of the equation, producing

$$x^4 + 2x^2y + y^2 = ax^2 + bx + c + 2x^2y + y^2.$$

On the left we have a perfect square $(x^2 + y)^2$. Call the expression on the right $q(x)$, and arrange the terms in decreasing powers of x to get

$$q(x) = (a + 2y)x^2 + bx + (c + y^2).$$

Ferrari's key insight was that this might also be a perfect square. If so, with a perfect square on each side, our quartic equation will be solvable. Can we make this happen? That is, can we find a number y for which $q(x)$ is a perfect square? That will occur if the discriminant $b^2 - 4(a + 2y)(c + y^2)$ equals 0.

An example will clarify this. Let $a = 7$, $b = 10$, and $c = 16$. Then $q(x) = (7 + 2y)x^2 + 10x + (16 + y^2)$, which we want to be a perfect square. Try $y = 1$, so $q(x) = 9x^2 + 10x + 17$. Is that a perfect square? If so, then it has just one root. But the quadratic formula gives $(-10 \pm i\sqrt{512})/18$ so there are two roots. This shows that $q(x)$ is not a perfect square for $y = 1$. But it also tells us what must be done. We need to change y so that the quadratic formula involves $\sqrt{0}$ instead of $\sqrt{512}$.

For any choice of y, what appears inside the squareroot is the discriminant of q, given in our example by $10^2 - 4(7 + 2y)(16 + y^2)$. We want this to be zero. And it will be zero if $y = -3$. (This can be verified by substitution. Never mind, for now, how it was discovered.) And with $y = -3$, $q(x)$ becomes $x^2 + 10x + 25$, which is evidently a perfect square.

The example reveals the pattern for the general case: $q(x) = (a+2y)x^2 + bx + (c+y^2)$ will be a perfect square when its discriminant is zero. Thus, we want y to satisfy

$$b^2 - 4(a + 2y)(c + y^2) = 0,$$

or in descending form,

$$8y^3 + 4ay^2 + 8cy - b^2 + 4ac = 0.$$

This is a cubic equation in y, so Cardano's method will give us at least one real solution. Using it, the equation

$$x^4 + 2x^2y + y^2 = (a + 2y)x^2 + bx + (c + y^2)$$

has a perfect square on each side, and thus has the form $w^2 = z^2$. That means we can simplify to the equation $w = \pm z$. This gives us two equations, each of which is quadratic in x, and so leads us to four roots.

Let us apply this method to one of Cardano's examples: $x^4 = 12x - 3$, in which $a = 0$, $b = 12$, and $c = -3$. Introducing y, add $2x^2y + y^2$ to both sides of the original equation, producing

$$x^4 + 2x^2y + y^2 = 12x - 3 + 2x^2y + y^2.$$

Rearranging both sides then gives us

$$(x^2 + y)^2 = 2yx^2 + 12x + (y^2 - 3),\tag{5}$$

with a perfect square on the left. We make the right-hand side a perfect square as well by requiring that

$$144 - 8y(y^2 - 3) = 0.$$

In standard form, this becomes

$$y^3 - 3y - 18 = 0,$$

and $y = 3$ is a solution. (Here, we were fortunate to find a cubic with so simple a root, no doubt the result of careful planning by Cardano. However, even if that were not the case, at least one y would be produced by Cardano's method for cubic equations.)

Substituting in (5), we find

$$(x^2 + 3)^2 = 6x^2 + 12x + 6.$$

As anticipated the right-hand side is a perfect square, so the equation becomes

$$(x^2 + 3)^2 = 6(x + 1)^2,$$

and that leads to

$$x^2 + 3 = \pm(x + 1)\sqrt{6}.$$

Thus we arrive at two quadratic equations

$$x^2 + \sqrt{6}x + 3 + \sqrt{6} = 0$$
$$x^2 - \sqrt{6}x + 3 - \sqrt{6} = 0.$$

Solving them gives us the four roots of the original quartic, namely

$$\frac{\sqrt{6} \pm \sqrt{4\sqrt{6} - 6}}{2} \quad \text{and} \quad \frac{-\sqrt{6} \pm i\sqrt{4\sqrt{6} + 6}}{2}.$$

Ferrari's method leads to four solutions for any quartic. The cubic equation in y will always have a real solution, which leads directly to a pair of quadratic equations in x. Looked at another way, once we have a value of y, we get a factorization of the original quartic into two quadratic factors. In the example we can go from

$$(x^2 + 3)^2 = 6(x + 1)^2$$

to

$$(x^2 + 3)^2 - 6(x + 1)^2 = 0,$$

and hence to

$$[(x^2 + 3) + \sqrt{6}(x + 1)][(x^2 + 3) - \sqrt{6}(x + 1)] = 0.$$

By revealing all the roots in this way, Ferrari's solution for the quartic is more direct than Cardano's solution of the cubic. This runs contrary to the expectation that the algebra should get progressively more complicated as the degree increases. Of course, you have to solve a cubic as part of the solution of a quartic, so the latter is not really simpler than the former. But that point aside, the direct factorization of the quartic does seem simpler than the factorization based on complex cube roots for the cubic. Similarly, the analysis of root permutations works out more simply for four roots than for three, confirming the impression that the quartic is somehow simpler than the cubic.

On first studying the methods of Cardano and Ferrari, one is struck by the lack of any unifying strategy. Both methods depend on clever algebraic tricks, but the tricks are completely unrelated. This suggests that other solutions might exist, based on different algebraic tricks, which has proven to be the case. Since the time of Cardano and Ferrari, many alternative approaches for solving cubic and quartic equations have been discovered. In the next section we will consider some, and observe that there does appear to be something inescapable in these solutions. In all of the cubic solutions, a common feature appears, and the same is true for the quartic. Why does this happen? Why do algebraic tricks that work so effectively for cubic and quartic equations fail for higher order equations? These are precisely the questions that inspired Lagrange to focus on symmetric functions of roots, and so to lay the foundations for key ideas in modern algebra, including group theory and Galois theory. At the end of this chapter, we will review some of Lagrange's ideas. For now, we proceed to alternate solutions of cubic and quartic equations.

4.3   Alternate Solutions for Cubics

For future reference, each solution in this section will be labeled with a descriptive phrase or with the name of its discoverer and the approximate date of discovery.

Viète, 1591.   Begin with the equation $x^3 + ax + b = 0$. Introduce the substitution $x = y - a/(3y)$. After expanding and simplifying, the original equation becomes $y^6 + by^3 - a^3/27 = 0$. This is a quadratic in y^3, and so tells us that

$$y^3 = \frac{-b \pm \sqrt{b^2 + 4a^3/27}}{2}.$$

Given these values for y^3, extracting a complex cube root provides a possible value of y. For each choice of y, we obtain a root for the original cubic from $x = y - a/(3y)$.

Although the algebra follows a different path for this solution than for Cardano's, the equation for y is the same as the equation in Cardano's solution for u, and both approaches lead to equivalent equations for x.

Euler, 1770.   Euler's solution is essentially identical to Cardano's except that he starts with the substitution $x = \sqrt[3]{s} + \sqrt[3]{t}$ rather than Cardano's $x = u + v$. The algebra is the same. However, Euler's notation is worth mentioning here because it so similar to what he used in the case of the quartic.

Cayley, 1877.   Cayley proposed a modified version of Cardano's solution for the equation $x^3 = ax + b$, substituting $u^2v + uv^2$ for x, rather than $u + v$. Cayley's substitution leads to two equations in u and v, but this time only u^3 and v^3 appear. The system is

$$u^3 + v^3 = 3b/a$$
$$u^3v^3 = a/3.$$

The right-hand side of the second equation is a bit simpler than in Cardano's approach, where in place of $a/3$ we would find $(a/3)^3$. There is another difference between the two methods. In Cardano's approach, once we define u as one of the possible cube roots of u^3, we are left with a single choice of v. But in Cayley's solution, we can choose v as any cube root of v^3, independent of the choice of u. Nevertheless, Cayley's solution to $x^3 = ax + b$ is the same as making a change of variables $x = y\sqrt[3]{a/3}$ and then applying Cardano's method.

Equality of Two Cubes.   The idea for this method is to recast the equation $x^3 = ax + b$ in the form $A(x + m)^3 = B(x + n)^3$. Then we can find a solution by taking a cube root of each side, leading to

$$x = \frac{B^{1/3}n - A^{1/3}m}{A^{1/3} - B^{1/3}}.$$

Thus, one root of the original cubic will be obtained as soon as we know A, B, m, and n.

   To find them, express $A(x + m)^3 - B(x + n)^3$ in descending form and equate coefficients with $x^3 - ax - b$. That gives the system

$$A - B = 1$$
$$3Am - 3Bn = 0$$
$$3Am^2 - 3Bn^2 = -a$$
$$Am^3 - Bn^3 = -b.$$

Using the first two equations, A and B can be found in terms of m and n. Substituting in the third and fourth equations leads eventually to

$$m + n = \frac{3b}{a}$$
$$mn = \frac{a}{3}.$$

   Once again we are led to equations for the sum and product of two unknowns, and hence to two solutions of a quadratic equation. This system is identical to the one that appears in Cayley's approach if we set $m = u^3$ and $n = v^3$. Further investigation proves that the algebraic expressions for the roots revealed by this approach are equivalent to those obtained in Cardano's approach.

   An interesting story accompanies this solution of the cubic, as retold in the autobiography of mathematician Mark Kac. See Sidebar 4.2.

Two Cube Completions.   Here is another approach using the idea of a perfect cube, due to Frink [55]. Rewrite the equation $x^3 + ax = b$ as

$$\frac{x^3}{8} + \frac{3x^3}{8} + \frac{ax}{2} = \frac{b}{2}.$$

Mark Kac and the Cubic

Mark Kac (1914–1984) was a leading mathematician who made pioneering contributions to the modern development of mathematical probability, and in particular its applications to statistical physics. His work in the latter is commemorated, in part, by the Feynman-Kac path integral, named after Kac and Richard Feynman. With Paul Erdős he introduced probabilistic methods in number theory. He was the author of several books, including popular and philosophical works in collaboration with figures such as Stanislaw Ulam and Gian-Carlo Rota. Kac won many awards, among them the Birkhoff prize (awarded jointly by the AMS and SIAM) and on two separate occasions the MAA's Chauvenet prize. One of the Chauvenet prizes was for the paper *Can One Hear the Shape of a Drum?* [73]. Among mathematicians, he may be best remembered in connection with that paper. Kac also was invited to deliver quite a few prestigious lectures, including SIAM's John von Neumann Lecture, the MAA's Hedrick Lectures, and the AMS's Gibbs Lecture.

Kac grew up in Poland. In his autobiography [74], he describes how an early fascination with cubic equations led him to become a mathematician. When he was 15, he recalls,

> …I became obsessed with the problem of solving cubic equations. Now, I knew the answer, which Cardan had published in 1545, but what I could not find was a derivation that satisfied my need for understanding. When I announced that I was going to write my own derivation, my father offered me a reward of five Polish zlotys (a large sum and no doubt the measure of his scepticism). I spent the days, and some of the nights, of that summer feverishly filling reams of paper with formulas. Never have I worked harder. Well, one morning, there it was — Cardan's formula on the page. My father paid up without a word, and that fall

In preparation for completing the cube, set $3x^3/8 + ax/2$ equal to $3xy^2/2$, which means that $y^2 = x^2/4 + a/3$. Then we obtain

$$\frac{x^3}{8} + \frac{3xy^2}{2} = \frac{b}{2}. \tag{6}$$

We will complete the cube on the left side of this equation in two ways. Let $R = 3x^2y/4 + y^3$. Then adding R to both sides produces

$$\left(\frac{x}{2} + y\right)^3 = \frac{b}{2} + R,$$

while subtracting R from both sides gives

$$\left(\frac{x}{2} - y\right)^3 = \frac{b}{2} - R.$$

Sidebar 4.2 (cont.)

Mark Kac and the Cubic (cont.)

Kac

my mathematics teacher submitted the manuscript to "Mlody Matematyk" (The Young Mathematician).... When my gymnasium principal, Mr Rusiecki, heard that I was to study engineering, he said, "No, you must study mathematics; you have clearly a gift for it".

The solution Kac discovered was the one presented here under the heading, *Equality of Two cubes.*

As Kac continues his story, he followed the advice of his principal, and as a result escaped sure destruction in World War II. Had it not been for the opportunities he found for study abroad in mathematics, and in particular, the fortuitous timing of his travels, he would undoubtedly have perished alongside his parents and brother at the hands of the Nazis.

Now we take cube roots to obtain the equations

$$\frac{x}{2} + y = \sqrt[3]{\frac{b}{2} + R} \tag{7}$$

$$\frac{x}{2} - y = \sqrt[3]{\frac{b}{2} - R}, \tag{8}$$

whose sum gives x in terms of R:

$$x = \sqrt[3]{\frac{b}{2} + R} + \sqrt[3]{\frac{b}{2} - R}. \tag{9}$$

To complete the solution, we need to express R in terms of the coefficients a and b. So multiply (7) and (8) to obtain

$$\frac{x^2}{4} - y^2 = \sqrt[3]{\frac{b^2}{4} - R^2},$$

and observe from the definition of y that the left side is $-a/3$. Therefore

$$\frac{-a^3}{27} = \frac{b^2}{4} - R^2$$

and

$$R = \sqrt{\frac{b^2}{4} + \frac{a^3}{27}}.$$

Substitution in (9) thus gives the solution

$$x = \sqrt[3]{\frac{b}{2} + \sqrt{\frac{b^2}{4} + \frac{a^3}{27}}} + \sqrt[3]{\frac{b}{2} - \sqrt{\frac{b^2}{4} + \frac{a^3}{27}}}.$$

This derivation seems to avoid the quadratic equation that arose in every other solution. However, when we obtain the value of R by taking a square root, that is equivalent to solving a quadratic equation, though a simple one. Moreover, it is essentially what we would encounter if we solved the quadratics in the earlier approaches by completing the square. And the solutions found here are in exactly the same form as those obtained using Cardano's solution.

Factorization Identities. The quadratic formula can be understood as a consequence of the identity $r^2 - s^2 = (r - s)(r + s)$. Any monic quadratic can be put into the form $(x - h)^2 - s^2$ by completing the square, and then the identity provides a decomposition into linear factors. This idea can be extended to cubic equations.

On the right-hand side of the quadratic identity, we can think of the $+$ and $-$ as representing square roots of unity, that is, ± 1. For cubics, there is a similar identity that involves the three cube roots of unity, 1, ω, and ω^2, where $\omega = (-1 + i\sqrt{3})/2$. It can be expressed in various forms, including

$$r^3 + s^3 + t^3 - 3rst = (r + s + t)(r + \omega s + \omega^2 t)(r + \omega^2 s + \omega t)$$

and

$$\omega(r^3 + s^3 + t^3) - 3\omega^2 rst = (\omega r + s + t)(r + \omega s + t)(r + s + \omega t).$$

These are equivalent, and it is a matter of taste which form is more memorable, or more closely resembles the identity for the quadratic case. While either can be used to solve a cubic equation, we will look at the first version.

First replace r, s, and t with x, $-u$, and $-v$:

$$x^3 - u^3 - v^3 - 3xuv = (x - u - v)(x - \omega u - \omega^2 v)(x - \omega^2 u - \omega v).$$

This can be used to factor any cubic in the form $x^3 + ax + b$ by making $-(u^3 + v^3) = b$ and $-3uv = a$. Then the roots are $u + v$, $\omega u + \omega^2 v$, and $\omega^2 u + \omega v$. Though different in concept, this method is algebraically identical to Cardano's approach.

Change of Variables in Symmetric Equations. In Chapter 3, we considered the idea of solving a cubic by direct inversion of the equations expressing the coefficients in terms of the roots. If the cubic is $x^3 + ax^2 + bx + c = 0$ and the roots are r, s, and t, the equations are

$$r + s + t = -a$$
$$rs + rt + st = b \tag{10}$$
$$rst = -c,$$

and all we have to do is find r, s, and t given a, b, and c.

This can be carried out if we make a change of variables using

$$r = u + v + w$$
$$s = u + \omega v + \omega^2 w \tag{11}$$
$$t = u + \omega^2 v + \omega w,$$

with ω a primitive cube root of unity as before. This will transform (10) into a system in u, v, and w. If we can determine values for u, v, and w, we will then be able to find r, s, and t.

So substitute the expressions on the right-hand side of (11) for r, s, and t in (10). The resulting system can be simplified, using the identities $\omega^3 = 1, 1 + \omega + \omega^2 = 0$, and their variants, to obtain

$$3u = -a$$
$$3u^2 - 3vw = b \tag{12}$$
$$u^3 + v^3 + w^3 - 3uvw = -c.$$

Use the first equation to eliminate u from the other two. What remains are equations involving vw and $v^3 + w^3$, essentially the same as we have seen in many of the other approaches. They can be put into a form that specifies values for the sum and product of v^3 and w^3, thus giving v and w as cube roots of the solutions of a quadratic equation. With u, v, and w thus determined, the roots r, s, and t are given by system (11).

How does this method compare to Cardano's? In Cardano's approach, we must take the preliminary step of eliminating the quadratic term of the cubic. Imposing the same assumption here amounts to making $u = 0$. But then system (12) reduces to

$$vw = -b/3$$
$$v^3 + w^3 = -c,$$

the very equations that arise in Cardano's method.

Matrix Algebra.   The preceding solution is intriguing. With a simple change of variables it permits direct inversion of the elementary symmetric functions. This approach has the advantage that it is conceptually transparent. Anyone might think of it. But the change of variables used seems to be unmotivated and mysterious. Even if you recognize the close links between the algebraic combinations we saw earlier and the definitions of u, v, and w, it is still not obvious that the change of variables will make system (10) solvable. It is legitimate to ask how anyone would find this change of variables.

One answer is provided by matrix algebra. The idea uses the fact that roots of polynomials can be realized as eigenvalues of matrices. Thus, a given matrix leads to both a set of roots (or eigenvalues) and the corresponding polynomial. This provides another alternate method for solving cubics and quartics. At its heart, this method corresponds to making a linear change of variables in the equations for the elementary symmetric functions. But in this setting the change of variables arises very naturally. These ideas are developed in detail in [92].

There are still other solutions for cubics. In Chapter 2 we considered an approach using not radicals, but the similarly defined curly root function. There are also geometric constructions (including the one mentioned earlier, attributed to Omar Khayyam), an approach using differential equations, and even a solution based on origami, or paper folding. But solutions that can be reduced to algebraic manipulation with radicals inevitably turn out to be equivalent to Cardano's method.

This is not just a trivial consequence of the fact that all of the methods have to produce the same roots, because those roots might conceivably appear in different forms. This is dramatically illustrated by Cardano's example $x^3 + 6x - 20 = 0$, where the real root 2 appears as $\sqrt[3]{10 + 6\sqrt{3}} + \sqrt[3]{10 - 6\sqrt{3}}$. Every other method we have seen for solving a cubic produces this root in exactly the same outlandish form. Apparently, there is only one way to skin a cubic, at least as far as algebraic manipulation is concerned. We turn next to solutions of the quartic, where we will observe a similar phenomenon.

4.4   Alternate Solutions for Quartics

Quartics can be solved by a variety of methods, several of which are analogs of cubic methods. As before, each method will be identified with a discoverer or a short descriptive phrase. Unless noted otherwise, we always assume that the equation to be solved is $p(x) = x^4 + ax^2 + bx + c = 0$.

Descartes, 1637.   Factor the quartic into a product of two quadratics. The fact that $p(x)$ has no cubic term implies that the linear coefficients of the quadratic factors must be equal in magnitude and opposite in sign. Therefore, the factorization we seek has the form $(x^2 + ux + v)(x^2 - ux + w)$. Expand this into descending form and equate the coefficients with those of p. That produces the equations

$$v + w - u^2 = a$$
$$wu - vu = b$$
$$vw = c.$$

The first two equations can be rewritten as

$$v + w = a + u^2$$
$$v - w = -b/u,$$

leading to expressions for v and w in terms of u. When they are substituted into the remaining equation, we obtain the equation in u alone

$$u^6 + 2au^4 + (a^2 - 4c)u^2 - b^2 = 0. \tag{13}$$

This is a cubic equation in u^2, and so is solvable.

Although this looks different from Ferrari's solution of the quartic, the cubic equations that arise in each method are closely related. Substituting $2y - a$ for u^2 in (13) (and adjusting for the opposite signs of the coefficients in the two approaches) reproduces Ferrari's cubic equation in y exactly. Though Descartes' method uses a different approach, it leads to an equivalent cubic equation, and factors the quartic into the same product of quadratic factors.

Euler, 1770. Euler's solution to the quartic is an extension of his solution of the cubic. He begins by assuming $x = \sqrt{r} + \sqrt{s} + \sqrt{t}$. He squares this, simplifies, squares a second time, and eventually obtains

$$x^4 - 2(r + s + t)x^2 - 8\sqrt{rst}\,x + (r + s + t)^2 - 4(rs + rt + st) = 0.$$

Now his idea is to make this match the original quartic by choosing r, s, and t appropriately. That way, we will know that $x = \sqrt{r} + \sqrt{s} + \sqrt{t}$ is one root of the original quartic. Equate the coefficients of the two equations to derive

$$-2(r + s + t) = a$$
$$-8\sqrt{rst} = b$$
$$(r + s + t)^2 - 4(rs + rt + st) = c.$$

In these equations Euler recognized the elementary symmetric functions for three variables. With $r + s + t = \sigma_1$, $rs + rt + st = \sigma_2$, and $rst = \sigma_3$, the system can be rewritten

$$\sigma_1 = -\frac{a}{2}$$
$$\sigma_3 = \frac{b^2}{64}$$
$$\sigma_1^2 - 4\sigma_2 = c.$$

To simplify further, use the first equation to eliminate σ_1 from the last equation, yielding

$$\sigma_1 = -\frac{a}{2}$$
$$\sigma_3 = \frac{b^2}{64}$$
$$\sigma_2 = \frac{a^2 - 4c}{16}.$$

These equations specify the elementary symmetric functions of r, s, and t in terms of the known coefficients a, b, and c. This is exactly analogous to the situation where we specify the sum and product of two variables. And just as the two-variable case tells us that the unknown variables are roots of a particular quadratic, so we can conclude here that r, s, and t are the roots of the cubic polynomial having $-\sigma_1$, σ_2, and $-\sigma_3$ as coefficients. That is, they are the roots of

$$g(x) = x^3 + \left(\frac{a}{2}\right)x^2 + \left(\frac{a^2 - 4c}{16}\right)x - \frac{b^2}{64}.$$

Using known methods, we can solve this cubic to find r, s, and t, and thus derive a root of the quartic we began with.

Here we have a third algebraic approach, different in form from the other two. Once again, the analysis depends on solving a cubic. And once again, it is essentially the same cubic. In fact, with the substitution $x = u^2/4$ Euler's cubic becomes Descartes' cubic.

Change of Variables in Symmetric Equations.   As for the cubic, a quartic equation can be solved by direct inversion of the symmetric functions, after making a suitable change of variables. If we label the roots of $x^4 + ax^3 + bx^2 + cx + d$ as q, r, s, and t, then the original system is

$$q + r + s + t = -a$$
$$qr + qs + qt + rs + rt + st = b$$
$$qrs + qrt + qst + rst = -c$$
$$qrst = d.$$

Substitute $q = u+v+w+z$, $r = u+v-w-z$, $s = u-v+w-z$, and $t = u-v-w+z$. The first equation then shows that $u = -a/4$, and we can use that to eliminate u from the remaining equations. After simplification, they become

$$v^2 + w^2 + z^2 = \frac{3a^2 - 8b}{16}$$

$$8vwz + a(v^2 + w^2 + z^2) = \frac{a^3 - 16c}{16}$$

$$(v^4 + w^4 + z^4) - 2(v^2 w^2 + v^2 z^2 + w^2 z^2) - \frac{a^2}{8}(v^2 + w^2 + z^2) - 2a(vwz) = d - \frac{a^4}{256}.$$

This system can be simplified by using symmetric functions of v^2, w^2, and z^2: $\sigma_1 = v^2 + w^2 + z^2$, $\sigma_2 = v^2 w^2 + v^2 z^2 + w^2 z^2$, and $\sigma_3 = (vwz)^2$. Also, $v^4 + w^4 + z^4 = \sigma_1^2 - 2\sigma_2$. Using these identities, the system becomes

$$\sigma_1 = \frac{3a^2 - 8b}{16}$$

$$8\sqrt{\sigma_3} + a\sigma_1 = \frac{a^3 - 16c}{16}$$

$$\sigma_1^2 - 4\sigma_2 - \frac{a^2}{8}\sigma_1 - 2a\sqrt{\sigma_3} = d - \frac{a^4}{256}.$$

Although this appears to be complicated, it is a triangular system. The first equation tells us the value of σ_1. Substituting it into the second equation determines the value of σ_3. Then, substituting both σ_1 and σ_3 into the third equation establishes the value of σ_2. Thus, we get each σ in terms of a, b, c, and d. From this point the solution proceeds as in Euler's analysis. With known expressions for each σ_j, it follows that v^2, w^2, and z^2 are roots of a cubic. The solutions of that cubic lead to v, w, and z, and we already know that $u = -a/4$. At last, the values of u, v, w, and z lead to corresponding values for the roots, q, r, s, and t of the original quartic.

To relate this approach to the ones already considered, we impose the assumption $a = 0$. Then the system of equations is readily solved, revealing

$$\sigma_1 = \frac{-b}{2}$$

$$\sigma_2 = \frac{b^2 - 4d}{16}$$

$$\sigma_3 = \frac{c^2}{64}.$$

Thus, v^2, w^2, and z^2 are roots of

$$x^3 + \left(\frac{b}{2}\right) x^2 + \left(\frac{b^2 - 4d}{16}\right) x - \frac{c^2}{64}.$$

This is the same as the polynomial $g(x)$ that appeared in Euler's solution, except that here the coefficients b, c, and d, respectively, play the roles of the coefficients a, b, and c from the earlier analysis.

Matrix Algebra. The remarks following the change of variables solution for the cubic also apply to the quartic. It is intriguing that a simple change of variables permits the solution of a quartic by directly inverting the elementary symmetric functions. But the lack of a rationale for defining the new variables detracts from the appeal of this approach. As mentioned before, matrix algebra provides an alternative viewpoint for solving cubics, and leads in a natural way to an appropriate change of variables. This matrix algebra viewpoint works in the same way for the quartic as well.

Looking back over all of the solutions to both cubics and quartics, several patterns stand out. For all of the cubic solutions it is necessary to solve an auxiliary quadratic equation. Likewise, each solution of the quartic depends on finding the roots of an auxiliary cubic equation. Moreover, all of the cubic solutions depend on the roots of the same or closely related quadratics. Likewise, all of the quartic solutions involve variations of the same cubic. These observations suggest that beneath the surface appearance of fortuitous algebraic gimmicks, there is an underlying structure dictating the form of the solution. Such an idea occurred to Lagrange, who analyzed this structure. Lagrange wanted to understand the successes in solving cubics and quartics, as well as the failure of all efforts to solve quintics. Although he did not settle the question of the quintic, he laid the foundation that permitted others to do so. In the next section we will see how Lagrange's ideas of symmetry and root permutation can be used to solve quartic equations.

4.5   Solving Quartics with Symmetry

As detailed in Sidebar 4.3, Lagrange and Vandermonde analyzed the significance of symmetry in solutions of cubic and quartic equations, working independently and essentially simultaneously around 1770. Reportedly, although they started with the same approach, Lagrange went further and his work had the greater impact on later developments. The following discussion is based loosely on his approach.

We have seen many methods for solving cubic and quartic equations. They all have in common the solution of an extra equation. We set out to find the roots of one polynomial

$p(x)$ (let's call it the *original* polynomial), but along the way we discover we have to find the roots of an *auxiliary*[1] polynomial $A(x)$. When p is a cubic, A is a quadratic; when p is a quartic, A is a cubic. And in both cases, the coefficients of the auxiliary polynomial are polynomial combinations of the coefficients of the original polynomial.

Lagrange set out to understand how and why an auxiliary polynomial arises. He used ideas related to the elementary symmetric functions that we considered in Chapter 3. For each polynomial the coefficients are symmetric functions of its roots. In addition, the co-efficients of A depend on the coefficients of p, while the roots of p depend on the roots of A.

To illuminate the power of Lagrange's ideas, we will depart from his path. Where he analyzed a known solution of the quartic, we will pretend to know no solution. Then, using Lagrange's insights, we will see how to manufacture an auxiliary polynomial, based not on an algebraic trick, but on the understanding of symmetry.

To begin, we introduce notation for the coefficients and the roots of the original and auxiliary polynomials. Let $p(x) = x^4 + ax^3 + bx^2 + cx + d$ with roots q, r, s, and t. We will assume that A is a cubic, writing $A(x) = x^3 + Ux^2 + Vx + W$, and denote the roots u, v, and w.

Now, what properties must an auxiliary polynomial have? First, it must be possible to obtain the roots of p from the roots of A. After all, that is the entire point of having an auxiliary polynomial. Second, its coefficients must be expressible as functions of the coef-ficients of p, and for simplicity, we will require that they be polynomial functions. At the same time, for both polynomials, the coefficients are known functions of the roots. All of these relationships are shown in Fig. 4.1.

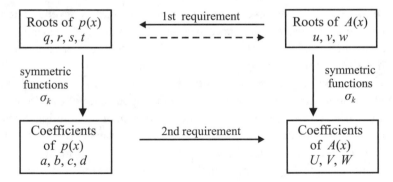

Figure 4.1. Related roots and coefficients for $p(x)$ and $A(x)$.

Inspired by these ideas, let us focus on how u, v, and w depend on q, r, s, and t, repre-sented in the figure by the dashed arrow. Given equations

$$u = f(q, r, s, t)$$
$$v = g(q, r, s, t) \qquad (14)$$
$$w = h(q, r, s, t)$$

we can satisfy the first requirement for an auxiliary polynomial by solving for q, r, s, and t in terms of u, v, and w. For the second requirement we make a key observation. The

---

[1] This is sometimes also referred to as a *resolvent* polynomial.

Lagrange and Vandermonde

Lagrange

Joseph-Louis Lagrange (1736–1813) and Alexandre-Théophile Vandermonde (1735–1796) made independent discoveries in the analysis of polynomial equations. A sketch of their contributions is given by Edwards [47, §15]. Edwards explains that although Vandermonde's early work in this area included promising insights, he (Vandermonde) went no further with it. The analyses of Vandermonde and Lagrange had much in common, but later developments primarily credited the contributions of Lagrange. Edwards' says

> Although Vandermonde had insightful ideas in other areas of mathematics as well, he does not seem to have followed these up either, and he is remembered today only because the name "Vandermonde determinant" was given to a determinant which, ironically, does not occur in his work at all.

> Unlike Vandermonde, who was French but did not have a French name, Lagrange had a French name but was not French. He was Italian (he was born with the name Lagrangia and his native city was Turin) and at the time of the publication of his [*Réflexions sur la Résolution Algébrique des Équations*, 1770–1771] he was a member of Frederick the Great's Academy in Berlin. When he left Berlin in 1787 he went to Paris, where he spent the rest of his life and where, of course, he was a leading member of the scientific community; this has tended to reinforce the impression that he was French. He was certainly the greatest mathematician of the generation between Euler and Gauss, and, indeed, has a secure place among the greatest mathematicians of all time.

Sidebar 4.3

coefficients of A will be expressible in terms of the coefficients of p if permutations of the roots of p leave the set $\{u, v, w\}$ unchanged.

To see why, note that (14) makes U, V, and W functions of q, r, s, and t, as suggested by the figure. Now consider what happens in (14) when we permute q, r, s, and t. We have not assumed that f, g, and h are symmetric functions, so the individual values of u, v, and w

may be altered. But this must amount to a permutation of u, v, and w if we assume the set $\{u, v, w\}$ remains unchanged. Then, U, V, and W will not change at all, since they depend symmetrically on u, v, and w. Thus U, V, and W are symmetric functions of q, r, s, and t, and so expressible in terms of the elementary symmetric functions σ_k by the fundamental theorem on symmetric polynomials (page 64). This shows that U, V, and W are expressible in terms of the coefficients of p, which differ from the σ_k by at most a change of sign.

To complete the construction of an auxiliary polynomial, we must specify the functions f, g, and h in (14). Here we follow Lagrange and define

$$u = (q + r)(s + t)$$
$$v = (q + s)(r + t) \tag{15}$$
$$w = (q + t)(r + s).$$

With these equations it is easy to check that every permution of q, r, s, and t leaves the set $\{u, v, w\}$ unchanged. Therefore, the coefficients of A will be expressible in terms of the the the coefficients of p. And this is not just a theoretical result. Knowing neither the roots of p nor those of A, we can find the coefficients of A explicitly as functions of the coefficients of p. This echoes our earlier work with symmetric combinations of roots, for example the sum of the squares, which we expressed in terms of the coefficients. In essence, when we are dealing with explicit symmetric functions of the roots, the fundamental theorem on elementary symmetric polynomials allows us to invert the mapping from roots to coefficients of a polynomial. Thus, in the figure, we can add an arrow from the coefficients to the roots of p. This provides a path from the coefficients of p to the coefficients of A, and so satisfies the second requirement for an auxiliary polynomial.

However, before carrying out this step, we should verify that it will lead to a solution of the original quartic p. Since A is a cubic, once we know its coefficients we can find its roots, u, v, and w. But will we then be able to solve (15) for q, r, s, and t?

From u, v, w to q, r, s, t. Alone, the three equations of (15) are not enough to find the four roots of p. But we also know all of the coefficients of p. From our knowledge of the elementary symmetric functions,

$$q + r + s + t = -a. \tag{16}$$

With this additional equation, (15) can be solved for q, r, s, and t, as we shall now see.

Using (16), we eliminate t from (15) to obtain a solvable system of three equations in three unknowns. The algebra is simplified if the original quartic has no cubic term. That can always be arranged, we know, by making a change of variables. So without loss of generality, we assume that $a = 0$. Then $t = -q - r - s$, and 15 becomes

$$-(q + r)^2 = u$$
$$-(q + s)^2 = v$$
$$-(r + s)^2 = w.$$

This leads to

$$q + r = \alpha$$

$$q + s = \beta$$

$$r + s = \gamma$$

where $\alpha = \pm\sqrt{-u}$, $\beta = \pm\sqrt{-v}$, and $\gamma = \pm\sqrt{-w}$. Thus we obtain a linear system in q, r, and s, with solution

$$q = \frac{1}{2}(\alpha + \beta - \gamma)$$

$$r = \frac{1}{2}(\gamma - \beta + \alpha)$$

$$s = \frac{1}{2}(\gamma + \beta - \alpha).$$

And since $t = -(q + r + s)$, we also get

$$t = -(\alpha + \beta + \gamma)/2.$$

This verifies that we can find the roots of p once we know the roots of A.

Where does that leave us? We have seen that, with an understanding of symmetry, it is possible to formulate proposed roots of an auxiliary polynomial in terms of the unknown roots of a quartic. Without actually computing either the auxiliary polynomial or its roots, we have deduced that the auxiliary coefficients will be computable in terms of the known original coefficients, and that the roots of the original polynomial will be obtainable from the auxiliary roots. In this way, it is possible to engineer a method for solving quartics, without actually working out the steps of the method in detail.

To complete this analysis we should carry out the construction of $A(x)$. As usual, we may assume that the original quartic has no cubic term, so that $q + r + s + t = 0$. However, even with this simplifying assumption, the algebra required to find $A(x)$ remains forbidding. First, we know how the coefficients U, V, and W depend on the roots u, v, and w:

$$U = -(u + v + w)$$

$$V = uv + vw + wu$$

$$W = -uvw.$$

Next, using (15), we determine U, V, and W as functions of q, r, s, and t. They will be symmetric functions, and expressing them in terms of the elementary symmetric functions is the final step. That will give U, V, and W in terms of a, b, c, and d.

Although this program can be completed by hand, modern technology provides an easier alternative. We can use a computer algebra system, as discussed in Sidebar 3.3. This is practical even without the simplifying assumption $q + r + s + t = 0$. Using Maple and trial and error, it took me about half an hour to express U, V, and W in terms of the elementary symmetric functions in q, r, s, and t. Here are the results:

$$U = -2\sigma_2$$
$$V = \sigma_2^2 + \sigma_1\sigma_3 - 4\sigma_4$$
$$W = -\sigma_1\sigma_2\sigma_3 + \sigma_3^2 + \sigma_4\sigma_1^2.$$

How does this approach compare with those discussed in the preceding section? To compare the auxiliary polynomial derived here with the ones found earlier, we again impose the constraint $q + r + s + t = 0$, or equivalently, $\sigma_1 = 0$. Then the formulas for the coefficients U, V, and W simplify to

$$U = -2\sigma_2$$
$$V = \sigma_2^2 - 4\sigma_4$$
$$W = \sigma_3^2.$$

If we write the original polynomial as $x^4 + ax^2 + bx + c$, then $a = \sigma_2$, $b = -\sigma_3$, and $c = \sigma_4$. With those substitutions, the auxiliary polynomial becomes

$$A(x) = x^3 + Ux^2 + Vx + W = x^3 - 2ax^2 + (a^2 - 4c)x + b^2.$$

This is closely related to the auxiliary polynomials we saw in the previous section. For example, the auxiliary polynomial featured in Descartes' method is $-A(-u^2)$.

This shows that in all of the quartic methods we have considered, the auxiliary polynomial is essentially the same as the one found by Lagrange's analysis. Lagrange knew that the solvability of the cubic and quartic depended on the existence of auxiliary polynomials, which he showed could be discovered using the tools of symmetry. Attacking the quintic equation in the same way, Lagrange was led to an auxiliary polynomial of sixth degree, and he could find no way to reduce it. Ultimately he gave up the attempt to solve higher degree equations, concluding, in the words of Kline [101, p. 605] that *either the problem was beyond human capacities or the nature of the expressions for the roots must be different from all those thus far known.*

This conclusion was later proven correct. Our final topic for this chapter is the insolvability by radicals of equations of degree 5 and higher.

4.6   Quintic and Higher Degree Equations

It has already been mentioned that no general methods exist for solving quintics and higher degree equations in terms of radicals. The proof depends on an analysis of permutations of roots, and is fully developed in what today is called Galois Theory. It is beyond the scope of this book to give anything like a complete account of this topic. But having come so far in the discussion of polynomial equations, it would be a shame not to at least describe the main ideas in general terms.

We begin with the idea of a permutation group. As an instance, we might consider a set of four objects, say $\{q, r, s, t\}$. A permutation rearranges their order. Formally, a permutation is a function that maps the set of elements to itself, creating a one-to-one correspondence. One example of such a pairing is shown in Fig. 4.2.

Since a permutation is one-to-one, it is invertible. Reversing the directions of all the arrows defines a permutation, and applying the original and reverse permutations in order has the same effect as the identity function, $f(x) = x$. In general, when two permutations are composed, that is, applied one after the other, the result is again a permutation. If we denote the first permutation by α and the second by β, then for any element x of our set, the result of the combined operation will be $\beta(\alpha(x))$. It is customary to denote the combined operation $\beta\alpha$.

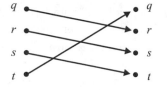

Figure 4.2. A sample permutation.

The set of permutations of any set forms an algebraic system called a *group*. Conceptually that means that the elements are invertible, and together they form a closed system. The same may be true of a subset of a group. In the set $\{q, r, s, t\}$ let α be the permutation that exchanges q and r, leaving s and t fixed. Then α is its own inverse, because applying it twice restores q and r to their original positions. The set consisting of α and the identity function thus constitute a closed system. It, too, is a permutation group, but not the group of all possible permutations of $\{q, r, s, t\}$, so it is called a *subgroup* of the full permutation group.

We have seen that the idea of permuting roots of a polynomial arises naturally when we express the roots in terms of radicals. In Cardano's solution of the cubic, there was an ambiguity in extracting a cube root of u. For complex u, there are three different cube roots, and switching from one to another permutes the roots of the original cubic. But we do not wish to deal only with polynomials that can be solved by radicals, so we need to understand how permutations might arise in other ways.

The modern viewpoint is to focus on number systems. We begin with the rational numbers, the system that contains the coefficients of our original polynomial. Next we extend the number system to include a root of our polynomial. If the root is, say, $3 + \sqrt{5}$, then we can form the set of numbers of the form $a + b\sqrt{5}$ with rational a and b. This formulation emphasizes $\sqrt{5}$ as a new element that must be incorporated into the number system. But we could equally well have incorporated the other square root of 5. In the end we arrive at the same number system. But the ambiguity in the choice of square roots of 5 has the effect of defining a transformation of the number system. The root $3 + 4\sqrt{5}$ defined as a result of one choice would instead have been $3 - 4\sqrt{5}$ had we made the other choice. This corresponds to the transformation $a + b\sqrt{5} \rightarrow a - b\sqrt{5}$, and is an analog of complex conjugation. There are two key properties of this transformation. First, it leaves all of the original elements of the number system unchanged. Second, it permutes the roots of our original polynomial. Thus, if $2 + 3\sqrt{5}$ is a root of a polynomial with rational coefficients, $2 - 3\sqrt{5}$ is also a root, and the transformation changes one into the other. More generally, permutations of roots arise any time we have a transformation of the extended number system that leaves elements of the original number system unchanged. The set of all transformations of this type gives rise to a group of permutations of the roots.

Galois theory is the study of number system transformations and the corresponding permutation groups. The root permutation group for a particular polynomial depends on the coefficients, and can either be the full permutation group or a subgroup. Distinguishing between these cases and understanding each group's structure are central aspects of the theory.

The idea of extending number systems also provides a way to consider roots that are expressed in terms of radicals. We begin again with the rational numbers, which we will call system 1. We can extend this to a larger number system, system 2, by incorporating one new element, $\sqrt[m]{a}$, where a is some element of system 1. Next we extend the number system again by incorporating a second new element $\sqrt[m']{b}$, with m' possibly different from m, and with b some element of system 2. Continuing, we can build up a number system that incorporates any radical expression through a chain of extensions. At each stage of the chain, we incorporate one new mth root into the preceding number system.

What we add at each stage is a root to a very simple polynomial, one of the form $x^m - r$. In fact, all of these are roots of one master polynomial, the product of all the individual simple polynomials. Because the factors are very simple, Galois theory makes it possible to analyze the root permutation groups for the master polynomial, and the transformations of number systems at each stage of the process. Out of this comes a key result: If a fifth degree polynomial has roots expressible in terms of radicals, then the root permutation group for this polynomial cannot be the full group of permutations of the five roots (and similarly for polynomials of degree greater than 5).

At this point, we have enough of the background to consider an example: $p(x) = 3x^5 - 15x + 5$. Methods of calculus show that this polynomial has three real roots and two non-real complex roots. Using Galois theory, that is enough to imply that the root permutation group for this $p(x)$ is the full group of permutations of five roots. But it was stated earlier that this could not occur for a polynomial whose roots are expressible in terms of radicals. This shows that the roots of $3x^5 - 15x + 5$ cannot be expressible in terms of radicals.

What is the significance of this example? It shows that there cannot be a general method for solving quintics akin to the ones for cubics and quartics. If there were such a method, it would have to apply to $3x^5 - 15x + 5$, giving the roots in terms of radicals. But we know the roots have no such expression.

That does not mean that *no* quintics can be solved in terms of radicals. For example, we saw how to solve palindromic quintics with radicals in Chapter 2. But at least some quintics are not solvable using radicals, and that is enough to rule out the existence of a general method.

This completes our exploration of methods for solving polynomial equations. As mentioned early on, the idea of searching for roots in terms of radicals is somewhat arbitrary, and reflects the historical development of the subject. But this tradition is completely natural in the context of the elementary mathematics curriculum. As we have seen, cubics and quartics, like quadratics, have readily understood solutions using only elementary algebra. On the other hand, there are no such general solutions for equations of higher degree. Not surprisingly, it is easier to exhibit methods for solving cubics and quartics using radicals than it is to prove that corresponding methods cannot exist for higher degree equations. This is a central result of modern mathematics, and the ideas on which it rests remain a cornerstone for ongoing research in the field. Here, we have traced the beginnings of Galois theory, from properties of elementary symmetric functions and permutations of roots of polynomials arising out of the consideration of solutions for cubics and quartics .

Even without understanding all of the details of Galois theory, it is worthwhile to have some idea of what it involves. It is Galois theory, after all, that provides the complete picture of solvability in terms of radicals: there are known radical methods for degree four or less, and no such methods are possible for quintics or higher degree polynomials. That is certainly worth knowing.

4.7   History, References, and Additional Reading

For general historical reading on polynomial equations, both Katz [94] and Kline [101] are recommended. A more focussed treatment is provided by Edwards [47]. This work carefully develops the ideas of Lagrange, Galois, and the other key figures in the search for roots of polynomial equations and provides a rich source of historical information as well. Stewart makes the development of methods to solve polynomial equations the central theme of his account of mathematical symmetry [153].

Galois' life makes a dramatic tale, and some of the retellings apparently have emphasized drama over accuracy. For a fascinating account of the true story and some of the exaggerations, see Rothman [136]. Peterson [129] presents a lighter overview of the Galois story.

The history of polynomial equations is tightly connected to the history of algebraic methods and notation, which is nicely summarized by Gouvea and Berlinghoff [15]. In a related vein, Kleiner [99] argues that it was precisely the investigation of Cardano's method for cubics that forced the introduction of complex numbers.

In the discussion of Cardano's solution, we encountered a surprising arithmetic fact:

$$2 = \sqrt[3]{10 + 6\sqrt{3}} + \sqrt[3]{10 - 6\sqrt{3}}.$$

There is an article on identities of this sort by Osler [128].

References for the alternate solutions of quartics and cubics are: Viète's solution of the cubic, [101, p. 269]; Euler's solution of the cubic and quartic, [42]; Descartes' solution of the quartic and Cayley's solution of the cubic, [8, p. 20–22]; equality of two cubes, [49, 80, 156]; two cube completions, [55]; factorization identities, [91]; change of variables, [127, 160]; matrix algebra, [92].

Solutions to cubic and quartic equations can be constructed using paper folding techniques. An explanation of the solution of cubics can be found in [69, activity 6]. For more in depth discussions of cubics and quartics see [2, 46].

There are a great many other papers about the solution of cubics and quartics in expository mathematics journals, including [3, 65, 163, 168, 170].

The biographical information on Mark Kac in Sidebar 4.2 was taken from [126]. I learned about Kac's solution of the cubic in Roy's paper [137]. Although Kac describes solving the cubic in his autobiography, he does not say how he did it. Roy gives the details of the derivation that Kac published as a student.

Part II

Maxministan

OR most tourists, a visit to Maxministan consists of a day or two spent in the capital, Optimopolis, usually as part of a package tour arranged through one of the many Calculusian travel agencies. The tours take in the standard sights: maximizing a differentiable real function on an open or closed interval, first and second derivative tests, remembering to check the endpoints, as well as colorful marketplaces featuring such traditional handicrafts as livestock pens bounded on a side by a river, boxes formed from rectangles by cutting squares from the corners, windows in the shape of a rectangle surmounted by a semicircle, and circular cylinders designed to contain a specified volume. Some travelers return a second time as part of a follow-up tour to the multi-variable districts, admiring the higher dimensional geometry of gradients, tangent planes, normal vectors, restrictions to boundaries, and, in the borough of constrained optimization, Lagrange multipliers.

For the more adventurous explorer, Maxministan offers a rich and varied landscape. The serious student of Maxmini lore can spend years surveying the realm, never finding an end to new and interesting discoveries. But even without such dedication, there is much to discover in the immediate environs of the capital. In the following chapters, we will investigate a few of the many opportunities that await just beyond the city's boundaries.

Our point of departure, a discussion of Lagrange multipliers, begins inside the capital proper. Almost immediately we find unfamiliar ground in the form of Lagrangian functions. This is an approach seldom seen these days, at least in mathematics classes, although it is closest to Lagrange's original inspiration. In this conception, an initial constrained optimization problem with objective function f is transformed into a related unconstrained optimization problem with objective function F, the Lagrangian function. Some in our party are sure to recognize this approach, remembering a well known heuristic justification for its validity. But few if any will suspect the fatal conceptual flaw lying hidden within this popular heuristic. We will see an alternative interpretation that is both correct and as intuitively compelling as the original, if not more so.

Our second expedition will be a combination of several short trips to minor attractions. Lagrange multipliers will return to play a significant role in several, including an examination of the idea of duality in optimization problems. This is different from duality in linear programming, and is inspired by the meaning of duality in the context of the isoperimetric problem. As we shall see, it is naturally expressed as a symmetry between objective and constraint functions in one of the vector formulations of Lagrange multipliers.

Succeeding excursions will look at some particular examples of max-min problems, with features distinguishing them from most standard exercises. We will see an unexpected geometric property of ellipses and the use of envelopes to solve the ladder problem.

5

Leveling with Lagrange:
Constrained Maxima and Minima
with Lagrangian Functions

Our tour of Maxministan begins at a popular spot in the standard curriculum, where Lagrange multipliers are used to solve constrained max/min problems. These are problems of the following type:

Max/min Problem: Find the point of the curve $xy = 1$ that is closest to the origin.

The distance from a point (x, y) of the curve to the origin is $D = \sqrt{x^2 + y^2}$, which we wish to minimize. It simplifies the problem to minimize D^2 instead, since the distance and its square are minimized at the same point. So we will minimize $f(x, y) = x^2 + y^2$. However, we are not free to choose just any point (x, y). We are permitted only to consider points on the *constraint curve*. If we define a *constraint function* $g(x, y) = xy - 1$, the problem can be restated in a standard form: Find the minimum value of $f(x, y)$ subject to the condition that $g(x, y) = 0$.

In the absence of a constraint, the typical approach to a max/min problem is to find points where the derivatives of f equal 0 (as well as considering the boundaries of the domain). This is not valid for a constrained problem, however. Because of the constraint, we can consider only a subset of points (x, y), which usually excludes the points where the derivatives vanish. Indeed, it would be unreasonable to expect to find a solution without taking into account the constraint function g. We need a more sophisticated approach than finding zeros of the derivative. It is provided by the Lagrange multipliers technique.

The standard approach to Lagrange multipliers, as it appears in most calculus books, calls for locating points (x, y) where the gradient vectors ∇f and ∇g are parallel. (A review of the ∇f notation and related topics is provided in Appendix C at the website for this book [87].) With the additional assumption that $\nabla g \neq 0$, the gradients of f and g are parallel when $\nabla f = \lambda \nabla g$ for some real multiplier λ. Combined with the equation $g = 0$, this gives necessary conditions for a solution to the constrained optimization problem. In more formal terms, these ideas are summarized in the following theorem. For concreteness it is stated for two variables, although it generalizes to n variables.

Lagrange Multiplier Theorem (LMT).   *Let f and g be continuously differentiable functions of two real variables, and let $S = \{(x, y)|g(x, y) = 0\}$. Suppose that the restriction of f to S assumes a local maximum or minimum at a point (x^*, y^*) where $\nabla g \neq 0$. Then there exists a real number λ^* such that $\nabla f(x^*, y^*) = \lambda^* \nabla g(x^*, y^*)$.*

The equations $\nabla f(x, y) = \lambda \nabla g(x, y)$ and $g(x, y) = 0$ are together called the Lagrange multiplier conditions for the problem of optimizing f subject to the constraint $g = 0$. Usually there are only finitely many solutions (x^*, y^*, λ^*). Then the finite set of pairs (x^*, y^*) necessarily includes all points at which the constrained objective function assumes local maxima and minima. Thus, the LMT applies to constrained optimization problems in the same way that the first derivative test applies to unconstrained problems.

There is an alternative to the standard approach to Lagrange multipliers that reflects the technique's historical development (see Sidebar 5.1) and is popular in books on mathematical methods for economics. Here is how it works. As before, we wish to maximize or minimize a function f subject to a constraint $g = 0$. Define a new function (called a *Lagrangian function*), $F = f + \lambda g$ that incorporates both the objective function and the constraint, and in which λ is now a new independent variable. Then, as if seeking the maximum or minimum of F without constraint, set all its partial derivatives, including $\partial F/\partial \lambda$, equal to zero. This gives us necessary conditions for a solution to the original constrained problem. It is the *Lagrangian function* approach to Lagrange multipliers.

Although both approaches produce the same necessary conditions and lead to identical solutions of constrained optimization problems, they seem to be quite different in concept. The standard approach can be understood in terms of the geometry of the objective and constraint functions, and necessary tangency conditions at the solution. There are several different ways to think about these conditions, each leading to an intuitive justification for the LMT. The Lagrangian function approach is usually described analytically rather than geometrically. It is often presented with a memorable intuitive explanation that is appealing, but unfortunately incorrect. At the end of the chapter we will see a geometric justification for the Lagrangian function approach that improves on the traditional one. I call this approach *Lagrangian leveling*.

5.1   Intuitive Justifications for the Standard Approach

The standard formulation of the LMT involves an objective function f and a constraint function g, defined over a domain of optimization in $\mathbb{R}^n$. The situation can be understood in several different ways. We can look at the graphs of f and g in $\mathbb{R}^{n+1}$, or at their level sets in $\mathbb{R}^n$. We can deal with them directly or parametrically. The constraint can be used to eliminate a variable, thus reducing the dimension of the domain, or combined with the objective function to increase the dimension of the range. For each approach, there is an intuitive geometric conception of why the LMT holds.

Let us consider several of these intuitive arguments. Individually, they are intrinsically interesting. Viewing them together expands our understanding of the LMT, and showcases the rich geometric context of the theorem. Because of the geometric emphasis, the point of each argument is to show that at the solution to the constrained optimization problem the gradients of the objective function and constraint function are parallel.

Tangent Level Curves. Suppose that at point p, f assumes a local maximum of M over a constraint curve S defined by the condition $g(x, y) = 0$. Then p lies on both S and the level set L where $f = M$. Usually, the level set where a function f has value c forms a boundary between the points of the domain where f exceeds c and those where it is less than c, at least locally. This must occur for a smooth function unless c is a local maximum or minimum of the function. We will assume that p is not the location of a local unconstrained maximum or minimum of f, since at such points, $\nabla f = 0$ and the Lagrange conditions are thus satisfied. Therefore, L will be a boundary between points in the domain where f is greater than M and points where f is less than M. The curve S cannot cross the boundary, because for all of the points on S near p, f must take a value that is less than or equal to M. So, since S and L touch at p, and since S cannot cross L, we conclude that the two sets are tangent at p. See Fig. 5.1.

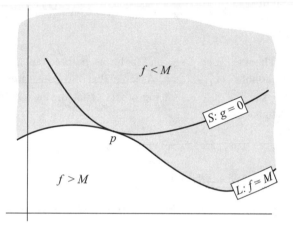

Figure 5.1. Tangency condition for level curves $f = M$ and $g = 0$.

The tangency of L and S at p shows that they share a common normal direction. Since ∇f is normal to L and ∇g is normal to S, they are parallel. That is the condition asserted by the LMT.

Parameterization. Suppose that at p there is a local maximum of f subject to the constraint $g = 0$, and let S be the set of all points satisfying the constraint. Consider a parameterization $\mathbf{r}(t)$ of a curve C in S through p. Then for some t^*, $\mathbf{r}(t^*) = p$.

Now observe that $f(\mathbf{r}(t))$ achieves a local maximum at t^*, so its derivative vanishes there. Using the chain rule, this shows that $\nabla f(p) \cdot \mathbf{r}'(t^*) = 0$. Therefore $\nabla f(p)$ is orthogonal to $\mathbf{r}'$. Also $\mathbf{r}'$ is tangent to S because it is tangent to the curve traced by $\mathbf{r}(t)$. Thus, $\nabla f(p)$ is orthogonal to any tangent vector of S at p, and hence to S itself. At the same time, ∇g is also orthogonal to S at p because S is a level set of g.

Except in degenerate cases, there is a unique direction orthogonal to the level set of a function. Thus, since $\nabla f(p)$ and $\nabla g(p)$ are both orthogonal to S, they must be parallel. See Fig. 5.2, which illustrates the three-variable case.

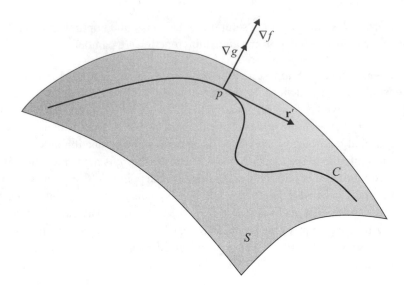

Figure 5.2. At a constrained extremum, both ∇f and ∇g are orthogonal to the velocity vector $\mathbf{r}'$ for any curve C on S.

Implicit Function Theory. Let f, g, and p be as before, but assume that f and g are functions of n variables, $x_1, \ldots, x_n$. Assume that the constraint $g = 0$ implicitly defines one variable, say x_n in terms of the others. That means that there is a function $y(x_1, \ldots, x_{n-1})$ such that

$$g(x_1, \ldots, x_{n-1}, y(x_1, \ldots, x_{n-1})) \equiv 0.$$

Otherwise expressed, there is a function

$$\Phi(x_1, \ldots, x_{n-1}) = (x_1, \ldots, x_{n-1}, y(x_1, \ldots, x_{n-1}))$$

such that the composite map $g \circ \Phi$ is identically zero.

Finding the maximum of f over the constraint set is equivalent to finding the unconstrained maximum of $f \circ \Phi$ over the domain of y. For two variables, $f \circ \Phi$ depends on a single variable as illustrated in Fig. 5.3.

For the unconstrained problem, at the solution $\nabla(f \circ \Phi)$ vanishes. Thus, the chain rule says

$$(\nabla f)^T \cdot d\Phi = 0,$$

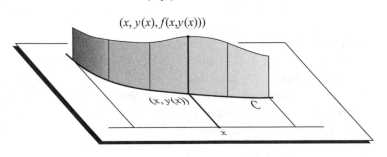

Figure 5.3. Graph of f with y implicitly defined.

where $d\Phi$, the matrix of partial derivatives of the component functions of Φ, is

$$d\Phi = \begin{bmatrix} 1 & 0 & 0 & \cdots & 0 \\ 0 & 1 & 0 & \cdots & 0 \\ 0 & 0 & 1 & \cdots & 0 \\ \vdots & \vdots & \vdots & & \vdots \\ 0 & 0 & 0 & \cdots & 1 \\ \frac{\partial y}{\partial x_1} & \frac{\partial y}{\partial x_2} & \frac{\partial y}{\partial x_3} & \cdots & \frac{\partial y}{\partial x_{n-1}} \end{bmatrix}.$$

Also, since $g \circ \Phi$ is identically zero, we have $\nabla(g \circ \Phi) = 0$, and the chain rule gives us

$$(\nabla g)^T \cdot d\Phi = 0.$$

Thus, ∇f and ∇g are both orthogonal to the columns of $d\Phi$. Because $d\Phi$ has $n - 1$ independent columns, ∇f and ∇g must be parallel.

Directional Derivatives. For two variables, we may visualize the domain of f as a subset of the x-y plane, and the graph of f as a surface in $\mathbb{R}^3$. The constraint set is then a curve C in the plane. The graph of the restriction of f to C, defined as $G_C = \{(p, f(p))| \ p \in C\}$, is a space curve lying on the full graph of f. Now suppose that f assumes a maximum over C at a point $p \in \mathbb{R}^2$. Then $(p, f(p))$ must be a high point of G_C, and so it has a horizontal tangent line T there.

In a neighborhood of p, the curve C is virtually identical to its tangent line through p. Once an orientation for the tangent line has been specified by the choice of a unit tangent vector $\mathbf{u}$, the restriction of f to the tangent line may be considered a function of a single real variable. Its derivative at p is called the directional derivative of f in the direction of $\mathbf{u}$, and denoted $D_{\mathbf{u}} f$. This must be zero, because as noted above, T is horizontal.

In general, $D_{\mathbf{u}} f$ can be computed as $\nabla f \cdot \mathbf{u}$. Since $D_{\mathbf{u}} f$ is zero at p, $\nabla f(p)$ is orthogonal to $\mathbf{u}$. But $\nabla g(p)$ is also orthogonal to $\mathbf{u}$, because $\mathbf{u}$ is a tangent vector to the curve $g = 0$. This shows that $\nabla f(p)$ and $\nabla g(p)$ are parallel.

The configuration is shown in Fig. 5.4. It does not extend easily to more than two variables, for which at least four dimensions would be required to visualize the graph of f.

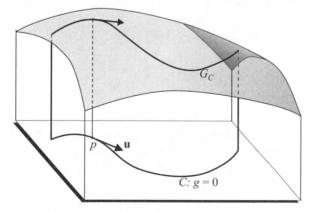

Figure 5.4. Directional derivative in the direction of the constraint curve vanishes at the constrained maximum.

Combined Mapping with f and g.   Next we consider a refreshingly different and strikingly beautiful geometric rationale for the LMT, credited to a 1935 work of Carathéodory in [131], where it is referred to as the *Carathéodory Multiplier Rule*. It applies most generally to the case of n variables and $n - 1$ constraints, but we will consider it here for $n = 2$. As before, we wish to maximize $f(x, y)$ subject to $g(x, y) = 0$. This time we combine the two functions to define a mapping $\Phi : \mathbb{R}^2 \to \mathbb{R}^2$,

$$\Phi(x, y) = (f(x, y), g(x, y))$$

(see Fig. 5.5). In the figure, we assume the axes in the image plane are labeled u and v, so that the mapping has the alternate definition

$$u = f(x, y)$$
$$v = g(x, y).$$

The constraint set $g = 0$ is precisely the set of points mapped by Φ to the u-axis. Therefore, maximizing f subject to $g = 0$ corresponds to finding the point on the u axis in the range of Φ that is as far to the right as possible. If the constrained optimization problem has a solution at (x^*, y^*), then $(u^*, 0) = \Phi(x^*, y^*)$ must be on the boundary of the range. Otherwise, it would be possible to go even further to the right. That is, if $(u^*, 0)$ is in the interior of the range, there must exist a point $(u, 0) = (f(x, y), g(x, y))$ with $u > u^*$. That would mean $f(x, y) > f(x^*, y^*)$ and $g(x, y) = 0$, contradicting the assumption that the constrained maximum of f occurs at (x^*, y^*).

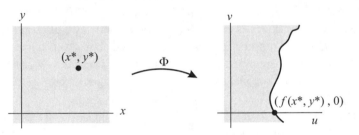

Figure 5.5. Action of the mapping Φ.

Next, consider the derivative of Φ, which can be expressed as the matrix

$$d\Phi(x, y) = \begin{bmatrix} \frac{\partial f}{\partial x} & \frac{\partial f}{\partial y} \\ \frac{\partial g}{\partial x} & \frac{\partial g}{\partial y} \end{bmatrix}.$$

Whenever $d\Phi(x, y)$ is invertible (or nonsingular), it is known from the theory of differentiable functions that Φ maps a neighborhood of (x, y) to a neighborhood of $\Phi(x, y)$. In particular, that would put $\Phi(x, y)$ in the interior of the range. Thus, since $\Phi(x^*, y^*)$ is on the boundary of the range, we conclude that $d\Phi(x^*, y^*)$ is singular. That means the matrix has dependent rows, so $\nabla f(x^*, y^*)$ and $\nabla g(x^*, y^*)$ are parallel.

Penalty Functions.   Finally, some mention should be made of the idea of penalty functions, as this approach comes closest in spirit to the Lagrangian approach. It is most natural to consider this idea in a minimization problem. Suppose we wish to minimize $f(x, y)$ subject to the condition $g(x, y) = 0$. We form the function $F(x, y) = f(x, y) + \sigma g(x, y)$, and seek an unconstrained minimum. Assuming for the moment that $g \geq 0$, we expect that at the minimum the contribution of the $\sigma g(x, y)$ term will be as small as possible. If we increase the value of σ and find a new minimum, the contribution of $\sigma g(x, y)$ should again be as small as possible, thus driving g to take on a smaller value. In effect, the $\sigma g(x, y)$ term imposes a penalty for allowing $g(x, y)$ to exceed zero, and the larger the value of σ, the greater the penalty. If we permit σ to increase without bound, in the limit the solution to the unconstrained minimization of F ought to approach the constraint $g = 0$. This can be used iteratively to seek numerical estimates for constrained minima. It can also be used to prove variants on the LMT (see [66, pp. 255-261] and [120]).

Excluding the penalty function approach, the various rationales presented above are visually persuasive. Each takes a slightly different point of view, and each geometrically justifies the LMT's conclusion that ∇f and ∇g must be parallel at the solution to the constrained optimization problem. They give us many ways of understanding why the standard approach to the LMT is valid.

Traditionally, there has been no comparable geometric justification for the Lagrangian function approach. However, it is frequently presented with an analytic justification in terms of incorporating the constraint into the objective function. This turns out to be a misconception. Why and how are the focus of the next section.

5.2 The Lagrangian Function Approach

Let us review in a bit more detail the two approaches to Lagrange multipliers. As in the introduction, let $f(x, y)$ and $g(x, y)$ be continuously differentiable, and suppose that we wish to find the maximum of f subject to the constraint $g = 0$. In the standard approach, the LMT says that if the solution occurs at a point (x, y) where $\nabla g \neq 0$, then there exists a real λ such that $g(x, y) = 0$ and $\nabla f(x, y) = \lambda \nabla g(x, y)$. Thus, the Lagrange conditions are

$$\frac{\partial f}{\partial x} = \lambda \frac{\partial g}{\partial x}$$

$$\frac{\partial f}{\partial y} = \lambda \frac{\partial g}{\partial y} \tag{1}$$

$$g(x, y) = 0.$$

In the Lagrangian function approach we define $F(x, y, \lambda) = f(x, y) + \lambda g(x, y)$, and regard F as a function of three variables. Following the traditional rationale for this approach, we observe that a necessary condition for F to achieve a local maximum is that all of its partial derivatives vanish. Thus we arrive at the system of equations

$$\frac{\partial f}{\partial x} + \lambda \frac{\partial g}{\partial x} = 0$$

$$\frac{\partial f}{\partial y} + \lambda \frac{\partial g}{\partial y} = 0 \tag{2}$$

$$g(x, y) = 0.$$

The last equation, which arises from $\frac{\partial F}{\partial \lambda} = 0$, assures us that any solution of the system satisfies the constraint equation.

These conditions differ from (1) only in the sign of λ. Thus, both approaches lead to the same conditions, and ultimately to the same solution. There is no suggestion in what follows that the Lagrangian approach leads to incorrect answers. Only the justification for the method is being questioned.

As formulated here, the Lagrangian function approach is based on an attractive, but incorrect, justification — that the problem of finding a maximum (say) of f subject to a constraint is transformed into a search for a maximum of F without constraint. The trouble is, the LMT says only that the solutions occur among the critical points of F, that is, among the points where $\nabla F = 0$. It gives no basis for suggesting that these will be points where F must have a maximum.

Unexpectedly, it turns out that a critical point of F satisfying the Lagrange conditions is *never* a local extremum. It is *never* the case that a constrained optimum of f can be obtained by finding an unconstrained optimum of the Lagrangian F. The contrary belief, though incorrect, is seductively attractive and frequently repeated. We will call it the *transformation fallacy*.

5.3   An Example

We now consider an example demonstrating the transformation fallacy. Our goal is to see that an extremum of a function f subject to a constraint need not correspond to an unconstrained local extremum of the Lagrangian function F. Indeed, any example will do, because the Lagrangian function essentially *never* possesses any local maxima or minima.

With that in mind, let us consider the perfectly pedestrian problem stated at the start of the chapter, to find the point of the curve $xy = 1$ that is closest to the origin. In the standard formulation, we must minimize $f(x, y) = x^2 + y^2$ subject to the constraint $g(x, y) = 0$ where $g(x, y) = xy - 1$. By symmetry, we may restrict our attention to (x, y) in the first quadrant. The Lagrange multiplier conditions (2) become

$$2x + \lambda y = 0$$
$$2y + \lambda x = 0$$
$$xy = 1.$$

The only solution is $(x^*, y^*, \lambda^*) = (1, 1, -2)$. Geometric considerations show that $(1, 1)$ is indeed the closest point of the curve $xy = 1$ to the origin, and $f(1, 1) = 2$ is the minimum value of f over the constraint curve.

Now we ask, is $(1, 1, -2)$ a local extremum of $F(x, y, \lambda) = x^2 + y^2 + \lambda(xy - 1)$? To see that it is not, it will be enough to show that the restriction of F to a particular plane through $(1, 1, -2)$ has a saddle point there. Then any neighborhood of $(1, 1, -2)$ includes points where F assumes both greater values and lesser values than $F(1, 1, -2)$, which can therefore be neither a local minimum nor a local maximum.

Consider the plane generated by unit vectors parallel to $\nabla g(x^*, y^*) = (1, 1)$ and the λ axis. In (x, y, λ) coordinates, the unit vectors are $\mathbf{u} = (1/\sqrt{2}, 1/\sqrt{2}, 0)$ and $\mathbf{v} = (0, 0, 1)$.

Therefore, a point on the plane may be expressed parametrically as

$$\mathbf{r}(s,t) = (1, 1, -2) + s\mathbf{u} + t\mathbf{v}$$
$$= (1 + s/\sqrt{2}, 1 + s/\sqrt{2}, t - 2).$$

Note that $(s, t) = (0, 0)$ corresponds to the point $(1, 1, -2)$, where ∇F is zero.

To restrict F to the plane, we define $h(s, t) = F(\mathbf{r}(s, t))$. We know that h will have a critical point at $(0, 0)$. To see whether this is a local extremum, we use the second derivative test (see Appendix C at the website for this book [87].) First, compute the composition $F(\mathbf{r}(s, t))$ to obtain

$$h(s, t) = (\sqrt{2}s + s^2/2)t + 2.$$

The Hessian matrix of second partial derivatives,

$$\begin{bmatrix} \frac{\partial^2 h}{\partial s^2} & \frac{\partial^2 h}{\partial s \partial t} \\ \frac{\partial^2 h}{\partial s \partial t} & \frac{\partial^2 h}{\partial t^2} \end{bmatrix} = \begin{bmatrix} t & s + \sqrt{2} \\ s + \sqrt{2} & 0 \end{bmatrix},$$

has determinant -2 at $(s, t) = (0, 0)$. Since it is negative, we know that $h(s, t)$ has a saddle point at the critical point $(0, 0)$. This shows that $F(x, y, \lambda)$ cannot have a local maximum or minimum at $(1, 1, -2)$. In fact, $(1, 1, -2)$ and $(-1, -1, -2)$ are the only critical points of F, which therefore has *no* local maxima or minima.

This reveals the flaw in the transformation fallacy, which tells us to solve the constrained optimization problem by finding the unconstrained minimum of F. That is impossible for no minimum exists. Nevertheless, the constrained problem has a solution.

5.4   A Theorem

What happened in the example always happens. For any objective function and constraint, the restriction of the Lagrangian function to a plane parallel to $\nabla g(x^*, y^*)$ and the λ axis has a saddle point at (x^*, y^*, λ^*), as shown in the following theorem. Although stated for functions of two variables, it extends naturally to functions of n variables.

First, we will establish a few notational conventions. Partial derivatives will be expressed using subscript notation, so f_x is the partial derivative of f with respect to x and f_{xy} is the second partial derivative with respect to x and y. We will abbreviate $\Phi(x^*, y^*)$ by Φ^* for any function Φ. In particular, $f^* = f(x^*, y^*)$, $\nabla f^* = \nabla f(x^*, y^*)$, and $f_{xx}^* = f_{xx}(x^*, y^*)$.

Theorem.   *Let f and g be functions of two variables, with continuous second derivatives. Let $F(x, y, \lambda) = f(x, y) + \lambda g(x, y)$. Then, if (x^*, y^*, λ^*) is a critical point of F at which ∇g^* is not zero, F has a saddle point at (x^*, y^*, λ^*).*

Proof.   At any critical point (x^*, y^*, λ^*), all of the partial derivatives of F vanish. This shows that $g^* = 0$ and $\nabla f^* + \lambda^* \nabla g^* = 0$. We also assume that ∇g^* is not zero.

Without loss of generality, we may assume $(x^*, y^*) = (0, 0)$ and ∇g^* points in the direction of the x axis: these conditions may be brought about by translating the x-y plane and rotating it about the λ-axis, neither of which alters the character of (x^*, y^*, λ^*) as a saddle point (or not). With these assumptions, $\nabla g^* = (g_x^*, g_y^*) = (a, 0)$ for some $a \neq 0$.

Next we will consider the restriction of F to the plane $y = 0$ in (x, y, λ) space. Let

$$h(x, \lambda) = F(x, 0, \lambda) = f(x, 0) + \lambda g(x, 0).$$

Then

$$\nabla h(x, \lambda) = (f_x(x, 0) + \lambda g_x(x, 0), g(x, 0)).$$

This vanishes at $(x, \lambda) = (0, \lambda^*)$, so $(0, \lambda^*)$ is a critical point of h.

We show that $(0, \lambda^*)$ is a saddle point of h by using the second derivative test for a function of two variables. We need the Hessian matrix of second partial derivatives, which is defined by

$$H(x, \lambda) = \begin{bmatrix} h_{xx} & h_{x\lambda} \\ h_{x\lambda} & h_{\lambda\lambda} \end{bmatrix} = \begin{bmatrix} f_{xx}(x, 0) + \lambda g_{xx}(x, 0) & g_x(x, 0) \\ g_x(x, 0) & 0 \end{bmatrix}.$$

At the critical point, the determinant of the Hessian is $\det H(0, \lambda^*) = -(g_x^*)^2$, and this is negative because we know $g_x^* \neq 0$. Therefore, the second derivative test shows that h has a saddle point at $(0, \lambda^*)$. Hence F must have a saddle point at (x^*, y^*, λ^*), as asserted.

∎

5.5   Presentations of the Fallacy

To many readers of this book, the Lagrangian function version of the LMT and the transformation fallacy will be familiar. But for others, it may hard to accept my statements about the fallacy, particularly given the preceding theorem. Can belief in the fallacy be as widespread as I have suggested?

It is a difficult question to answer. In casual conversation with colleagues, I have found that most mathematicians who remember Lagrange multipliers in any detail also are aware of the Lagrangian function version and have some recollection of the transformation fallacy. But almost none of them teach Lagrange multipliers in this fashion, and calculus texts generally follow the standard approach.

On the other hand, it is easy to find statements of the transformation fallacy in textbooks, particularly in mathematical economics. Some are explicit, like this description of Lagrange multipliers in an economics text:

> We now set out the *Lagrange multiplier technique* ... We proceed by introducing a new variable, λ, the **Lagrange multiplier,** and by forming the **Lagrange function** or **Lagrangean**
> $$\mathcal{L}(x_1, x_2, \lambda) = f(x_1, x_2) + \lambda g(x_1, x_2)$$
> We then carry out the unconstrained maximization of $\mathcal{L}$ with respect to x_1, x_2, and λ. [67, p. 588]

Other authors only guide students in the direction of the transformation fallacy, stopping short of an unambiguous assertion of the fallacy as fact. For example, Chiang [29, p. 373] says

> The essence of the Lagrange-multiplier method is to convert a constrained-extremum problem into a form such that the first-order condition of the free-extremum problem can still be applied.

This is technically correct in that it says only that the (necessary) condition will be applied, not that the point found will be an extremum of the Lagrangian function. And later the method is described as a matter of "screening the stationary values of [F] taken as a *free* function of three choice variables." That too is a fair description of the Lagrange method. On the other hand, Chiang also says

> ...with the constraint out of the way, we only have to seek the *free* maximum of [F], in lieu of the constrained maximum of [f].

This sounds perilously close to the transformation fallacy. Yet, even this statement is technically correct if interpreted to mean only that *seeking* the free maximum will turn up the constrained maximum, whether or not the latter actually is an instance of the former. The wording might have been carefully chosen to avoid saying anything technically incorrect, while still conveying the suggestion of maximizing F.

Economists are not the only authors who write this way. In a popular calculus text written by mathematicians ([68, p. 749]) Lagrange multipliers are first introduced using the standard approach. Later, the formulation with the Lagrangian $\mathcal{L}$ is presented, and it is shown that the earlier conditions are equivalent to the partial derivatives of $\mathcal{L}$ vanishing. The presentation concludes,

> In other words, (x_0, y_0, λ_0) is a critical point for the unconstrained problem of optimization of the Lagrangian $\mathcal{L}(x, y, \lambda)$. Thus the Lagrangian enables us to convert a constrained optimization problem to an unconstrained problem.

Here again, the wording avoids the explicit error of the transformation fallacy. But what other impression can the student gain from that final sentence? At any rate, the authors make no effort to steer students away from a seductive (and incorrect) interpretation.

Based on these examples, at least some texts in current use include the transformation fallacy explicitly or implicitly. At a guess, the vast majority of students who learn about Lagrange multipliers do so in calculus classes and via the standard approach. They are safe from the fallacy. For students of mathematical economics the risk of exposure is probably greater, but I have no idea how high the risk might be. Suffice it to say, the transformation fallacy issue is not on anyone's list of critical educational problems.

My interest is more academic. I say the perpetuation of a flawed intuitive explanation of the Lagrangian approach is a shame, not least because it represents a dilution of the power of a true intuition. And we have so many beautiful alternative intuitions to offer! On the other hand, the Lagrangian version of the LMT has its attractions. Why shouldn't it have a valid intuitive justification of its own? But of course it does, and we are just about ready to see it. First though, we consider a different question. Why does the transformation fallacy persist?

5.6   In Defense of the Fallacy

We have seen that a constrained maximum of f essentially never arises as an unconstrained maximum of F. Why does the transformation fallacy persist? Perhaps it is so attractive that it is rarely questioned.

In one way, it does seem very plausible. We get to change a constrained problem into an unconstrained problem, and that makes the problem easier. But we have to introduce an

What did Lagrange say about the multiplier method?

We have already encountered Lagrange's work on solving polynomial equations in Sidebar 4.3, and on polynomial interpolation (see page 33). But to calculus students and teachers, his name is probably most familiar in connection with his multiplier technique. In Lagrange's original development of this technique, it is the Lagrangian function approach that appears, although in a form where the multipliers are not considered as variables of the function. What I have called the standard approach is a later development.

Lagrange did not initially formulate the multiplier method for constrained optimization, but rather in analyzing equilibria of systems of particles. He reported this application in *Mécanique Analytique*, published in 1788 and available now in English translation [104]. Using series expansions to analyze the effects of perturbation at the point of equilibrium, Lagrange derived conditions on the first order differentials for systems under general assumptions. When the particle motions are subject to external constraints, he pointed out that the constraints can be used to eliminate some of the variables that appear in his differential equations, before deriving the conditions for equilibrium. He went on to observe that

> ...the same results will be obtained if the different equations of [constraint], each multiplied by an undetermined coefficient are simply added to [the general formula of equilibrium]. Then, if the sum of all the terms that are multiplied by the same differential are put equal to zero, as many particular equations as there are differentials will be obtained [104, p. 60].

Next, he explained how this procedure can be "...treated as an ordinary equation of maxima and minima," having carefully pointed out earlier that "the equation of a differential set equal to zero does not always represent a maximum or minimum" [104, p. 55].

Sidebar 5.1

additional variable, and that makes the problem harder. Trading off one effect against the other embues the Lagrange multiplier technique with a certain amount of *street cred*.

Moreover, this view mirrors the substitution method, in which we use the constraint equation to eliminate a variable. There, too, we transform a constrained problem into an unconstrained problem, and we also see a similar tradeoff between aspects that increase and decrease the difficulty of the problem. The difficulty is decreased by reducing the number of variables. It is increased by the algebraic complexity that arises when we eliminate a variable.

The persistence of the transformation fallacy may also depend on what appears to be a valid proof. The argument goes something like this. Because $\frac{\partial F}{\partial \lambda} = g$, a maximum of F will have to occur at a point where $g = 0$. Thus, maximizing F implicitly imposes the constraint condition. At the same time, if $F(x^*, y^*, \lambda^*)$ is a local maximum of F, then it is greater than or equal to the values assumed by F at nearby points where $g = 0$. So, since

What did Lagrange say about the multiplier method? (cont.)

Thus, Lagrange must have intended to convey that the conditions derived by his perturbation analysis could be found by formulating a Lagrangian function and proceeding as if seeking a maximum or minimum. But there is no suggestion that the equilibrium point must actually correspond to a maximum or minimum in general, nor that the existence of an extremum is necessary to establish the validity of the equilibrium conditions.

Having developed the multiplier technique in the analysis of equilibria, Lagrange proceeded to use it in other settings, notably in the calculus of variations [54]. Today's familiar application to constrained optimization problems was presented in two pages in his *Théorie des Fonctions Analytiques* of 1797, which appeared nearly a decade after his initial work with statics. As in the earlier work, Lagrange first established the conditions for a constrained extremum using series expansions. Then he pointed out that the same conditions can be obtained from a general principle [105, p. 198]. In translation, that principle is:

> When a function of several variables must be a maximum or a minimum, and there are one or more equations relating these variables, it will suffice to add to the proposed function the functions that must be equal to zero, multiplied each one by an indeterminate quantity, and next to find the maximum or minimum as if these variables were independent; the equations that one will find combined with the given equations will serve to determine all the unknowns.

On casual inspection, this appears to be a statement of the transformation fallacy. However, two points are significant. First, in saying to find the extremum as if the variables were independent, Lagrange clearly does not consider them to include the multipliers. This is reflected not only by the context in which he uses the phrase *these variables,* but also by his inclusion of the phrase *combined with the given equations.*

Sidebar 5.1 (cont.)

F and f are identical for the points where $g = 0$, we have found a local maximum of f among such points.

This is correct, but it suffers from two major flaws. First, it turns out that trying to maximize F is destined to fail — F will not have any local maxima. Consequently, the strategy of solving the original problem by finding the maximum of F loses much of its appeal: *You wanna find the constrained max of f? Well, just square this circle and duplicate this cube, and then your answer will be evident.*

The second flaw is more subtle, and has to do with the way Lagrange multipliers are used. The application is a two-stage process. First we find all solutions to the Lagrange conditions (as well as points where the hypotheses of the LMT fail, and possible locations of extrema on the boundary of the domain of f), and then choose from among them the point that solves our original problem. The strategy works because of our preknowledge that the solution is one of the points we have found. That is, aside from possible boundary solutions and singularities of the constraint, we must know that the Lagrange conditions

What did Lagrange say about the multiplier method? (cont.)

If the multipliers were considered as variables, there would be no need to mention the constraint equations separately, for they would appear as partial derivatives with respect to the multipliers. Second, it seems evident that Lagrange advised proceeding only *as if* seeking a maximum or minimum, and that the key point is the assertion of the final clause: the variables can be found by solving the given set of equations. Remember, too, that the validity of that assertion, established in a separate argument, did not depend on the existence of an unconstrained extremum at the solution point.

Apparently the idea of considering the multipliers as new variables for the optimization problem originated elsewhere, probably when someone recognized that the partial derivatives with respect to the multipliers are exactly the quantities that the constraints hold at zero. To a reader with that observation in mind however, it is easy to imagine that the quoted passage above, taken out of context, might encourage a belief in the transformation fallacy.

will be satisfied by the solution of the original problem, so that the Lagrange equations constitute a *necessary condition* for solutions to the constrained optimization problem.

The correct way to justify a necessary condition is to assume that a solution to the original problem is given, and then show that it satisfies the necessary conditions. This shows that the rationale above for the transformation fallacy is exactly backward. It begins with a point that satisfies the presumed necessary conditions, and argues that such a point is a local constrained maximum of f. Even if there were such points (and there aren't), there would be no assurance that they include *all* local constrained maxima of f. In particular, there is no assurance that maximizing F will reveal the global constrained maximum of f, which is what we wish to find.

In practice, the Lagrangian function approach has *nothing* to do with finding a local maximum of F. Rather, *all* the critical points of F are found, and these are further examined to find the constrained maximum of f. Along the way, no one ever checks to see which critical points are local maxima of F. Indeed, if textbook authors performed such a check in their examples of the Lagrangian technique, they would discover that critical points of F are never maxima.

The idea of maximizing F appears only as part of the transformation fallacy, and only it seems to justify the idea that $\nabla F = 0$ is a necessary condition for the desired constrained maximum of f. We have now seen that these ideas are flawed. As a justification for the Lagrangian technique, they do not stand up to scrutiny.

The logical confusion of necessity with sufficiency is enough to discredit the rationale for the transformation fallacy. However, knowing that F has no local maxima crystalizes the issue dramatically. It shows that a proper justification must explain why the critical points of F (and in fact, really why the *saddle points* of F) are significant. And it must do so without assuming that F is maximized somewhere. The transformation fallacy does not provide such a justification. Later we will see an argument that does.

Do authors believe the transformation fallacy? It is hard to tell. In virtually every case (excepting only where ∇g vanishes at the answer), the Lagrangian function fails to have local maxima and minima. In examples, we can easily detect this by computing at the solution point the eigenvalues of the Hessian matrix for $F(x, y, \lambda)$. This is all but effortless using modern mathematical software packages. So it is not a matter of counterexamples being hard to come by.

Perhaps some authors accepted the fallacy when they were students, and never questioned it. Others likely know quite well that solutions to the constrained optimization problem can only be associated with critical points (not necessarily local maxima or minima) of the Lagrangian, but choose to gloss over this point. It is enough, they might argue, if the students can remember and apply the Lagrange multiplier technique. No need to confuse them with epistemological technicalities. I disagree.

One reason for studying mathematics is to cultivate careful habits of thought. We should be skeptical of imprecisely formulated or understood intuitions, no matter how appealing or credible they appear. For the transformation fallacy, it may be plausible that absorbing the constraint into the objective function justifies passage to an unconstrained optimization. But as a matter of reflex, we should ask *why*? How exactly does that work? Teachers should encourage this reflex in their students, and be prepared to answer the questions it provokes. That would be impossible for the transformation paradox.

Heuristic arguments can be an effective way to communicate core ideas, but they must be correct, and they should guide us to correct proofs. Seeing how the formal proof reflects the intuitive core strengthens our understanding of both the proof and the intuition. The justifications in Section 5.1 do this, the transformation fallacy does not. Perpetuating it is wrong on two counts. First, it encourages students to believe something that is not true. Second, it deprives them of an opportunity to witness the interplay between a powerful heuristic and a valid proof.

5.7   Leveling the Playing Field

A new justification is needed for the Lagrangian function approach to the LMT. As we have seen, the traditional heuristic explanation is conceptually incorrect. The earlier justifications in terms of tangency conditions set a high standard. Each provides a persuasive intuition supporting the LMT, and one that can be refined into a proof. Is there a rationale for the Lagrangian function approach that meets this standard?

I claim that there is. It requires a change in how we look at the Lagrangian function. The transformation fallacy says that the Lagrangian function is a device that transforms a constrained problem into an unconstrained problem. An equally memorable idea, and one that is correct, is that the Lagrangian approach *levels the playing field*.

Let us consider once again the fundamental problem, to maximize $f(x, y)$ subject to $g(x, y) = 0$. Suppose that the solution occurs at (x^*, y^*). If we are lucky, this will be a local maximum of f disregarding the constraint. Then there is a high point on the graph of f at (x^*, y^*), and the tangent plane is horizontal. That is why we can find such points by setting the partial derivatives equal to 0.

Usually, though, this is not what happens. Visualizing the constraint curve as in Fig. 5.4, the maximum occurs at a high point of a particular curve G_C on the graph of f. The tangent plane is not horizontal there. Traveling along the plane in the direction of G_C we would experience a slope of 0, because the curve has a high point, but by moving off of the curve we can go even higher. In particular, traveling on the tangent plane perpendicular to G_C the slope is not 0.

But what if there were a way to remedy this defect — a way to deform the graph of f and level out the tangent plane? In the process, we do not want to disturb G_C, because that might change the location of the constrained maximum. But it should be all right to alter the values of the function at points that are not on the constraint curve. Near the high point of G_C, we can imagine pivoting the graph of f around the curve's horizontal tangent line. If we pivot by just the right amount, the tangent plane will become horizontal, and so detectable by setting partial derivatives to zero. That is the image of leveling the playing field.

Approaching this idea analytically, the simplest way to alter the values of the function f is to add a perturbation. But we want the amount we add to be 0 along the constraint curve, so as not to alter f for those points. We know that g is 0 along the constraint curve, so adding g to f is the sort of perturbation we need. It leaves the values of f unchanged along the constraint curve, but modifies the values away from that curve. Moreover, we can add any multiple of g to f and achieve the same effect. With this motivation, let us define a perturbed function

$$F(x, y) = f(x, y) + \lambda g(x, y).$$

We do not think of F as a function of three variables. Rather, we have in mind a family of two-variable functions, one for each value of λ. This is shown in Fig. 5.6. The surface with $\lambda = 0$ is the unperturbed graph of f. The other surfaces are graphs of different members of the family of F functions. They all intersect in the graph of f over the constraint curve, so the constrained maximum is the high point on the intersection curve.

The figure shows that choosing different values of λ imposes the pivoting action described earlier. Intuition suggests (and further analysis proves) that there is a choice of λ that makes the tangent plane horizontal. That is, if (x^*, y^*) is the location of the maximum value of f subject to the constraint $g = 0$, then for some λ^* it will be the case that $\nabla F(x^*, y^*) = 0$, and therefore we will have

$$\frac{\partial f}{\partial x}(x^*, y^*) + \lambda^* \frac{\partial g}{\partial x}(x^*, y^*) = 0$$

$$\frac{\partial f}{\partial y}(x^*, y^*) + \lambda^* \frac{\partial g}{\partial y}(x^*, y^*) = 0$$

$$g(x^*, y^*) = 0.$$

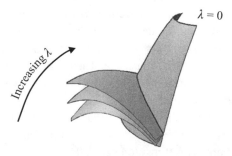

Figure 5.6. Graphs of several members of the family of F functions.

By finding every possible triple (x^*, y^*, λ^*) for which these equations hold, we obtain a candidate set that contains the solution to the constrained optimization problem. That is what the LMT says.

The existence of λ^* is easy to establish. At the maximum point over the constraint curve ($g = 0$ in Fig. 5.4), we know that the directional derivative $D_\mathbf{u} f$ in the direction of the curve is zero. And we assume that the derivative $D_\mathbf{n} f$ in the direction normal to the constraint curve is not 0. The idea of leveling is to make the derivative in the direction of $\mathbf{n}$ vanish for the perturbed function F. That is, we want

$$D_\mathbf{n} F = D_\mathbf{n} f + \lambda D_\mathbf{n} g = 0.$$

The normal vector $\mathbf{n}$ can be defined as the unit vector in the direction of ∇g. Then $D_\mathbf{n} g = \nabla g \cdot \mathbf{n} = |\nabla g|$, which can be 0 only if $\nabla g = 0$. Otherwise, for $\lambda^* = -D_\mathbf{n} f / D_\mathbf{n} g$, $D_\mathbf{n} F = 0$ and the modified function F will have a horizontal tangent plane.

From this analysis we can also rederive an interpretation of λ^* popular in the optimization literature for applications in economics. The ratio $D_\mathbf{n} f / D_\mathbf{n} g$ is the rate of change of the objective function f relative to a change in the constraint function g. It shows, for a given perturbation of the point (x^*, y^*) orthogonally away from the constraint curve, how the change in f compares to the change in g. The economists interpret this as the marginal change in the objective function relative to the constraint. It indicates, to first order, how the maximum value will change under relaxation or tightening of the constraint.

Leveling the graph of f using the Lagrangian function F offers an attractive way to think about Lagrange multipliers. It gives a clear and convincing intuitive justification of the Lagrangian approach. It also makes the existence of the necessary λ^* transparent, while leading naturally to the interpretation of λ^* as the marginal rate of change of the objective function relative to the constraint.

For those whose primary introduction to Lagrange multipliers has been recent calculus texts, familiarity with this topic may be limited to the standard approach, along with one or more of the geometric rationales described earlier. The Lagrangian function approach offers this audience a new insight about the technique, and helps place it in a historical context. The idea of leveling the playing field provides an additional rationale for the multiplier method. This is particularly interesting because it is so different from the rationales based on tangency conditions, and because it fits so nicely with the Lagrangian function formulation. And it is certainly superior to perpetuating the transformation fallacy.

5.8   History, References, and Additional Reading

There is a vast literature on optimization. It covers in great depth ideas that have been barely touched upon here, and far more. A good general reference most closely related to the topics discussed in this chapter is Hestenes [66]. It provides a general discussion of the LMT, including the transformation of constrained to unconstrained problems through a technique called augmentation and a detailed account of penalty functions. Although Hestenes does not use the terminology of leveling, the idea is implicit in his treatment of augmentability. Penalty functions are also explored in [120].

Pourciau [131] (winner of a Ford award) is highly recommended as a general account of multiplier rules, and has many references. Hassell and Rees [64] and Spring [152] both discuss second derivative tests for determining when a solution to the Lagrange conditions corresponds to a maximum, minimum, or saddle point for the constrained objective function f. Shutler [146] discusses this as well, and also the behavior of the Lagrangian function, considered as a function of both the original variables and the multiplier, at a critical point. The findings in that paper are similar to our discussion showing that critical points of F are generally saddle points.

As mentioned frequently, the Lagrangian function approach is a popular one in works on mathematical applications for economics. Examples of such works are [5, 29, 67]).

6

A Maxmini Miscellany

Optimization problems, where the goal is to determine how large or small something can be, are Maxministan's main claim to fame, and rightly so. But it is curious how often other sorts of mathematical results, with no obvious connection to optimization, can be established by solving max/min problems. In this chapter we will see several examples. We will also get a glimpse of some other elements of Maxmini culture, hinting at the mathematical diversity and richness of this realm.

We start with a well-known property of ellipses, that a tangent line at any point P makes equal angles with the lines from P to the foci. This property can be derived by solving a max/min problem in two different ways.

6.1 Milkmaids and Elliptical Mirrors

The max/min problem that we will look at is sometimes called the *Milkmaid* problem. The milkmaid must go from her cottage to the river, and then to the barn with a bucket of water for her cow. In Fig. 6.1, she starts at A, the river is indicated by line L, and the barn is at B. We want to find the point P on the river for which the milkmaid's journey (from A to P to B) is as short as possible. The mathematical formulation is to find a point P on line L so that the distance $AP + PB$ is a minimum.

We will solve the milkmaid problem in two ways. The first solution is elegantly simple. Reflect B across line L to determine B'. Draw a line from A to B'. Where it intersects L is the solution P to the milkmaid problem. See Fig. 6.2. The justification is that any path from

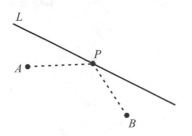

Figure 6.1. The milkmaid problem: find the shortest path from A to line L to B.

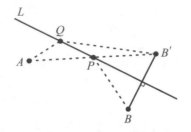

Figure 6.2. For any point Q on line L, the paths AQB and AQB' are equal in length. Since path APB' is shorter than AQB' for any $Q \neq P$, the minimal path from A to L to B must go through point P.

A to L to B corresponds to a path of equal length from A to L to B'. Thus, the minimal length path from A to B corresponds to the minimal length path from A to B', and that is clearly the straight line path.

From this solution, we obtain an interesting corollary. At the solution point P, the lines AP and PB make equal angles with L. To see why, note that AP and PB' make equal angles with L by the equality of vertical angles. Likewise, PB' and PB make equal angles with L because it divides triangle PBB' into congruent right triangles. This shows that AP and PB make equal angles with L. And the logic works both ways: if we take a point Q on the line for which AQ and QB make equal angles with L, we find that AQ and QB' also make equal angles with L. This shows that A, Q, and B' are collinear and hence that Q is actually P. Thus we have derived a characterization of the solution P to the milkmaid problem: it is the unique point from which the lines to A and B make equal angles with L.

This completes the first solution to the milkmaid problem. For the second solution we formulate the problem as a case of constrained optimization. Defining (x, y) as the coordinates of P, the objective is to minimize

$$f(x, y) = \text{the sum of the distances from } (x, y) \text{ to } A \text{ and } B$$

for (x, y) on line L. Applying Lagrange multipliers, at the solution point (x^*, y^*) the constraint curve (here, the line L) is tangent to a level curve S for f (see page 103). If the minimum value of f is D, then S consists of all points (x, y) (disregarding the constraint) for which $f(x, y) = D$. In other words, S is the set of points P such that $AP + PB = D$. This makes S an ellipse, defined as the locus of points for which the sum of the distances to two fixed points is constant (see Fig. 6.3). The fixed points are called the foci of the ellipse. In the present context, we have discovered that the solution to the milkmaid problem is a point of tangency between line L and an ellipse with foci at A and B.

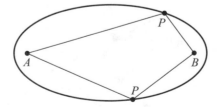

Figure 6.3. The distance from A to P to B is the same for every point P on the ellipse.

The two solutions to the milkmaid problem combine to establish a property of ellipses. Turning the situation inside out, we begin, not with the milkmaid problem, but with a point P on an ellipse E, whose foci are A and B. Let L be the tangent line to E through P. Now consider the milkmaid problem for A, B, and L. The solution (x^*, y^*) is a point of tangency between L and an ellipse with foci A and B. But there can be only one such ellipse tangent to L, so it must be E. And there can be only one point of tangency between E and L, namely P. Thus P is the solution to the milkmaid problem. Since our earlier characterization shows that from this solution the lines to A and B must make equal angles with L, we have derived the following conclusion.

Optical Property of an Ellipse: At any point P of an ellipse, the tangent line makes equal angles with the lines from P to the foci of the ellipse.

The optical property of ellipses is a familiar part of the study of conic sections. With calculus and analytic geometry, we can find the equations of the ellipse, tangent, and focal lines, and so determine the angles between the tangent and focal lines. But verifying the optical property in this way is surprisingly complicated. In contrast, the proof we have just seen is wonderfully simple. It is a fitting first example of deriving a mathematical result from a max/min problem.

6.2   Getting an Angle on Rotated Conics

Ellipses and the other conic sections are a traditional topic of study in analytical geometry. One of the most important results in this subject is the identification of the conic sections as graphs of quadratic equations. A general quadratic equation in two variables has the form

$$Ax^2 + 2Bxy + Cy^2 + Dx + Ey + F = 0 \tag{1}$$

where the quantities A through F are constants. Such an equation generally represents an ellipse, hyperbola, or parabola although in degenerate cases it can be an isolated point, one or more lines, or the empty set. Aside from the degenerate cases, the graphs always have one or two axes of symmetry, which, when $B = 0$, are horizontal or vertical. Then the equations can always be reduced to simple standard forms, which permit easy identification of the type of curve and the location of key geometric features. Conversely, an equation in one of the standard forms can always be put into the form of (1) with $B = 0$.

For example, consider

$$9x^2 + 4y^2 + 18x - 16y = 11.$$

We group the x terms together and the y terms together, and then complete the square for each, obtaining

$$9x^2 + 18x + 9 + 4y^2 - 16y + 16 = 11 + 9 + 16$$

or

$$9(x + 1)^2 + 4(y - 2)^2 = 36.$$

Dividing by 36 produces the standard form

$$\frac{(x + 1)^2}{4} + \frac{(y - 2)^2}{9} = 1.$$

Sidebar 6.1

Eavesdropping with the Optical Property

Where does the term *optical property* come from? It is a reminder that the equal angles condition characterizes the path of a beam of light undergoing reflection. When a beam of light is reflected, the incoming and outgoing rays make equal angles with the reflecting surface. Thus, if we consider the inside of the ellipse to be a mirror, then a ray of light from a focus to any point P of the ellipse will reflect along a ray making an equal angle with the ellipse's tangent at P. The optical property shows that this reflected path will pass through the other focus. If we put a light source at one focus that sends out rays in all directions, the light rays will reflect off the ellipse and reconverge at the other focus, as illustrated in Fig. 6.4. This is the optical meaning of the word *focus*. And since each ray of light will travel the same distance, rays emitted at the same time will reconverge at the same time.

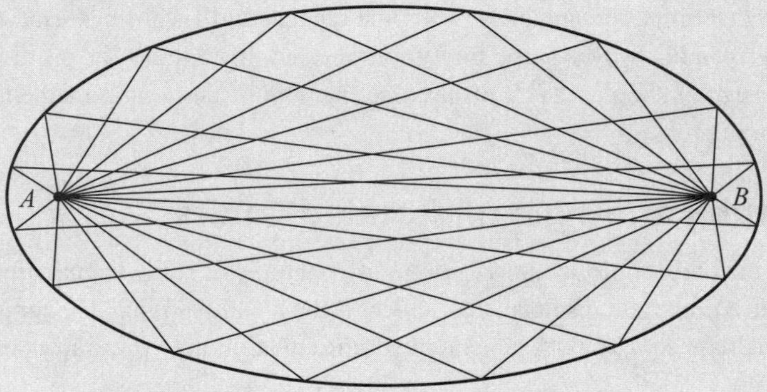

Figure 6.4. Optical Property of an Ellipse.
All light rays emanating from focus A reconverge at focus B.

This tells us that the curve is an ellipse centered at $(-1, 2)$, extending $\sqrt{4} = 2$ units left and right and $\sqrt{9} = 3$ units up and down from the center. See Fig. 6.5.

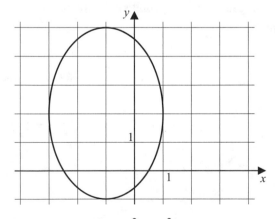

Figure 6.5. Graph of $9x^2 + 4y^2 + 18x - 16y = 11$.

Sidebar 6.1 (cont.)

Eavesdropping with the Optical Property (cont.)

But it would make just as much sense to refer to this as an acoustical property, because sound waves obey the same principle of reflection. Legend has it that our sixth president, John Quincy Adams, took advantage of it when he served in Congress.

Though it may be apocryphal, here is the story as it is told by tour guides at the U. S. Capitol. The House of Representatives used to meet in what is now Statuary Hall. Because the hall is elliptical in shape with an ellipsoidal domed ceiling, sounds generated at one focus reconverge at the other. Consequently, a whispered conversation near one focus could be overheard quite clearly by a listener at the other.

When John Quincy Adams served in the House of Representatives he put his desk at one of the foci. Pretending to be asleep at his desk, he often listened in on the conversations of the opposition who caucused near the other focus. Thus he turned a geometrical property of ellipses to his own practical advantage.

Statuary Hall

When an equation has a nonzero xy term, it represents a rotated conic section. It will still be an ellipse, parabola, or hyperbola, but its axes will no longer be horizontal or vertical. It is possible to define a new coordinate system whose axes are parallel to those of the conic section. This corresponds to introducing new variables, say u and v, in the equation for the curve. The transformed equation will have no uv term, and then we can use the methods of the $B = 0$ case.

Given an equation, how do we determine the correct angle of rotation? One common approach is to transform the equation with arbitrary rotation angle θ and then choose θ so that the coefficient of uv vanishes. Though more or less routine algebra, this can be tedious. Solving a particular max/min problem provides an alternative.

It turns out that the rotation angle depends only on the coefficients A, B, and C. So, if we can find the proper angle of rotation for an equation of the form $Ax^2 + 2Bxy + Cy^2 + F = 0$, then we can apply it to the general case.

In this special case, I claim that the rotation angle can be determined by finding the points where the curve is closest to the origin. If the curve is an ellipse, then it will be centered at the origin, to which the closest points are the ends of the minor axis. Finding those points reveals the axis, and hence the rotation angle. The hyperbolic case is similar because the closest points to the origin will be the vertices. For a parabola the situation is a little more complicated, so let us leave that aside for the moment. Our task, then, is to find the closest points to the origin on a curve with equation $Ax^2 + 2Bxy + Cy^2 + F = 0$. This is ideally suited for the Lagrange multipliers method.

As an example, consider the equation

$$5x^2 - 11xy + 3y^2 - 1 = 0. \tag{2}$$

Denoting the left-hand side of this equation by $g(x, y)$, let us seek the minima of $f(x, y) = x^2 + y^2$ subject to the constraint $g = 0$. We compute

$$\nabla f(x, y) = (2x, 2y)$$

and

$$\nabla g(x, y) = (10x - 11y, -11x + 6y).$$

At a point where f achieves a local minimum, the gradient vectors must be parallel. Since plane vectors (a, b) and (c, d) are parallel just when $ad = bc$, we have

$$2x(6y - 11x) = 2y(10x - 11y),$$

which simplifies to

$$11x^2 + 4xy - 11y^2 = 0. \tag{3}$$

The solutions of the max/min problem satisfy (2) and (3). But our goal is not to find the solutions. They are of interest only because they lie on an axis L of our curve. Our goal is to find the angle θ between L and one of the standard axes. To determine θ we will find the slope m of L. We can assume m exists because otherwise the axis is vertical and that cannot happen with a nonzero xy term in (2).

Axis L passes through the origin and a solution point P of the max/min problem. Since L has slope m, P must be of the form (t, mt). We also know that P satisfies the Lagrange conditions, (2) and (3). Substituting in the second we obtain

$$(11 + 4m - 11m^2)t^2 = 0, \tag{4}$$

while the first tells us that $t \neq 0$. So we have a quadratic equation for the slope of L:

$$m^2 - \frac{4}{11}m - 1 = 0.$$

Once we find m, the rotation angle is given by $\tan \theta = m$.

Although the equation has two roots, that causes no ambiguity. The constant term, -1, is the product of the roots (as we saw in Chapter 3) because the quadratic is monic. Therefore, the two roots are negative reciprocals of one another, corresponding to slopes of perpendicular lines. These will be the slopes of the x and y axes after rotating our coordinate system. Whichever m we choose, the rotated axes will align with the axes of our curve. The quadratic formula gives one root as

$$m = \frac{2 + \sqrt{125}}{11}$$

and the rotation angle is

$$\theta = \tan^{-1}\left(\frac{2 + \sqrt{125}}{11}\right). \tag{5}$$

It is interesting that we never finished solving the max/min problem. All we needed was the Lagrange multiplier conditions, which allowed us to find m. We did not solve (2) and (3), nor did we determine the points of the curve closest to the origin. Even though we now see that there is no need to do so, why not finish the problem just for fun?

Any solution to the Lagrange conditions must be of the form (t, mt) where $m = (2 \pm \sqrt{125})/11$. To find t substitute (t, mt) in (2), producing

$$t^2(5 - 11m + 3m^2) = 1.$$

We must consider both values of m in order to find all possible solutions to the Lagrange conditions. Even without explicitly solving the equation, we can predict what solutions will occur. If $a = 5 - 11m + 3m^2$ is positive there will be two values for t, corresponding to two solutions to the Lagrange conditions. Let us approximate a numerically for each m. Using a calculator, I find the values of m to be 1.1982 and -0.8346 to four decimal places. For the first value, $a = -3.8732$. This does not lead to a real value for t, so the first m does not produce any solutions to the Lagrange conditions. This shows that there are no points of intersection of the curve and a line through the origin with slope 1.1982. For the second value of m, calculation shows that $a = 16.2699$, leading to two values of t, and hence to two solutions of the Lagrange conditions.

On the basis of this evidence, we can infer that the given curve cannot be an ellipse. Otherwise, there would be two points of the curve closest to the origin and two furthest away, and all four of these points would satisfy the Lagrange conditions. Since we only found two solutions to the Lagrange conditions, the given curve must be a parabola or a hyperbola. Thus, the max/min analysis tells us not only the rotation angle θ, but also partial information about the type of curve.

If the same steps are applied to $Ax^2 + 2Bxy + Cy^2 = 1$, we get

$$m^2 + \frac{A - C}{B}m - 1 = 0.$$

Since $m = \tan\theta$, we see that

$$\tan^2\theta + \frac{A - C}{B}\tan\theta - 1 = 0.$$

We could now solve for θ. But it turns out to be better to isolate the constant $(A - C)/B$. That leads to

$$\frac{A - C}{B} = \frac{1 - \tan^2\theta}{\tan\theta}. \tag{6}$$

If you know your double angle identities, you may recognize the right-hand side of this equation. In fact, since

$$\tan 2\theta = \frac{2\tan\theta}{1 - \tan^2\theta}$$

(6) can be written

$$\tan 2\theta = \frac{2B}{A - C}.$$

Thus, we derive the general formula

$$\theta = \frac{1}{2}\tan^{-1}\left(\frac{2B}{A-C}\right).$$

A complete analysis of the properties of rotated conics extends beyond this. It is useful to know that a quadratic equation represents an ellipse, parabola, or hyperbola according to whether $B^2 - AC$ is negative, zero, or positive. I have not found a simple way to derive this fact from the max/min problem considered here. Even for finding the rotation angle the discussion is not complete, for we have not considered the parabolic case. While a similar analysis still applies, an equation of the form $Ax^2 + 2Bxy + Cy^2 = 1$ can represent only a degenerate parabola, consisting of two parallel lines. In this case, the axis of the parabola is parallel to the lines, and the solutions to the max/min problem lie on a line perpendicular to them. Thus the max/min approach will succeed in finding θ, but a complete justification gets somewhat involved. For these reasons, the max/min approach cannot be considered to be a substitute for the standard treatments of this subject. On the other hand, it is interesting how simply the Lagrange conditions lead us to the formula for the rotation angle θ.

6.3   Tracing the Borders of String Art Designs

Our next example of a result established by solving a max/min problem concerns string art designs. An example is shown in Fig. 6.6. Set nails or pegs every half a unit along the positive x and y axes, with strings stretched between the nails. As shown in the figure, a string stretches from 1 on the y axis to 5 on the x axis, from 2 on the y axis to 4 on the x axis, from 3 on the y axis to 3 on the x axis, and so on. The pattern of the strings seems to create a curve. But what curve is it? How can the equation of the curve be found?

Although to the eye the lines seem to make a smooth curve, the border is actually a polygonal path composed of short line segments. Additional lines can be added between the ones shown that will make the border even closer to a smooth curve. In the limit, if we consider a region Ω made up off *all* possible lines, we expect its boundary to be a smooth curve. These ideas can be made mathematically precise, and there is a procedure

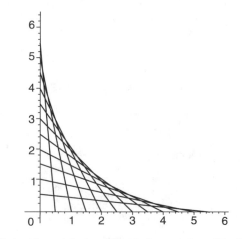

Figure 6.6. A string art design. What is the equation of the border curve?

for finding the equation of the boundary. We will see that it can be derived by solving a particular max/min problem.

First, let us clarify the idea of *all possible lines*. As shown in Fig. 6.6, the original string art design is made up of line segments whose x and y intercepts sum to six. In the diagram, the intercepts take on discrete values $1/2, 1, 3/2$, etc. To allow for all possible lines, let the x intercept a take on any real value from 0 to 6. Then, with y intercept $b = 6 - a$, we obtain the line with equation

$$\frac{x}{a} + \frac{y}{6-a} = 1. \tag{7}$$

If we were to graph the lines for every value of a from 0 to 6, they would completely fill a region of the plane. The intersection of that region with the first quadrant defines Ω. By definition, Ω is bounded on two sides by the coordinate axes. The remainder of the boundary is given by a smooth curve. We want to find the equation of that curve.

What we have here is a family of curves. For any numerical value of a, (7) represents a particular curve, and by varying a we generate a set of curves. Here the curves are all straight lines, but we might equally well consider families of circles, parabolas, or any other sort of curves. In general, suppose $F(x, y, a)$ is a smooth function of three variables and for each value of a, $F(x, y, a) = 0$ defines a curve C_a. Then F defines the family of curves $\{C_a | a \in \mathbb{R}\}$. We will refer to this as the family defined by F.

The curve that makes the boundary in the string art example is also known as the *envelope* of the family of lines. Envelopes and boundary curves are defined differently, but they are the same for many families of curves. Under some modest assumptions, a point (x, y) is on the envelope of the family defined by F if, for some a,

$$F(x, y, a) = 0 \quad \text{and} \quad \frac{\partial F}{\partial a}(x, y, a) = 0.$$

If the two equations can be combined to eliminate the parameter a, the resulting equation in x and y defines the envelope curve. This will be referred to as the *envelope algorithm*.

The subject of envelopes and their connection to boundary curves will be taken up in Chapter 10. Here, we will only derive the envelope algorithm conditions for the string art design and similar families of curves. In keeping with our recurring theme, the derivation makes use of a max/min problem.

It will simplify things to revise the equation defining our family of lines. To be consistent with the general description of the envelope algorithm, (7) should be put into the form

$$\frac{x}{a} + \frac{y}{6-a} - 1 = 0$$

But since we will be differentiating with respect to a, let us we rewrite it as

$$(6 - a)x + ay - a(6 - a) = 0. \tag{8}$$

This is one of the two equations for the envelope. To obtain the second equation we differentiate with respect to a, finding

$$-x + y - 6 + 2a = 0.$$

Can we eliminate a from these equations? Certainly. From the second

$$2a = x - y + 6.$$

Next, multiply both sides of (8) by 4, distributed as follows:

$$2(12 - 2a)x + 2(2a)y - 2a(12 - 2a) = 0.$$

Now substitute $x - y + 6$ for $2a$ and simplify:

$$2(12 - x + y - 6)x + 2(x - y + 6)y - (x - y + 6)(12 - x + y - 6) = 0$$
$$12x - 2x^2 + 2xy + 2xy - 2y^2 + 12y + x^2 - 2xy + y^2 - 36 = 0$$
$$-x^2 + 2xy - y^2 + 12x + 12y - 36 = 0$$
$$x^2 - 2xy + y^2 - 12x - 12y + 36 = 0.$$

This is the equation of the boundary curve for the string art pattern in Fig. 6.6.

We have reached our goal, observing the use of the envelope algorithm to find the equation of a boundary curve. Now let us see what the equation can tell us. It is quadratic and has a nonzero xy term, so it represents a rotated conic section. Using the methods of the preceding section, we can find that the rotation angle is 45° and the curve is a parabola.

We can verify this without having to rotate the axes. Set up the string art pattern with pegs on the lines $y = \pm x$ rather than on the x and y axes. With pegs at $(-6, 6), (-5, 5), \ldots,$ $(-1, 1)$, the origin, and at $(1, 1), (2, 2), \ldots, (6, 6)$, draw lines as in the original pattern. Then a generic line will connect $(-a, a)$ to $(6 - a, 6 - a)$, and expressing its equation in the form $F(x, y, a) = 0$ gives us a family of lines. Applying the envelope algorithm, the boundary curve is readily found to be a parabola with a vertical axis. This exercise is recommended for readers who wish to practice the application of the envelope finding process.

The appearance of a parabolic boundary curve in this example is a special case of a more general pattern. Suppose that the pegs for a string art design are equally spaced along two straight lines, intersecting at a point A. Let the end points for successive strings move toward A on one line and away from A on the other. For such a pattern, the border curve is always a parabola.

The envelope algorithm has a succinct description: differentiate the original equation with respect to the parameter, and combine with the original equation to eliminate the parameter. But why does this work? How is it derived? At last we come to a max/min problem.

In the original example we wish to find the boundary curve for the region Ω, as illustrated in Fig. 6.7. For a particular x, we can see visually how to find the corresponding boundary point. Start in Ω and move vertically as far as possible. That is, maximize y while holding x fixed.

To formalize this idea, take a point (x, y) in Ω. It must be a point of one of the curves in the family, so there exists an a such that $F(x, y, a) = 0$. Thus, the segment in Ω corresponding to a fixed x is found by varying y and a subject to the requirement $F(x, y, a) = 0$. For the points on the segment we want to determine the largest possible y.

This is a constrained optimization problem, with objective function $f(y, a) = y$ and constraint $g(y, a) = F(x, y, a)$, and where we think of x as fixed. Among all the points (y, a) for which $g = 0$, we wish to find the point with the maximal value of f. That is, we are to maximize f subject to $g = 0$.

Let us apply the Lagrange multiplier technique. At the solution point, the gradients of f and g must be parallel. Here, $\nabla f(y, a) = (1, 0)$, and $\nabla g(y, a) = (\partial F/\partial y, \partial F/\partial a)$. These

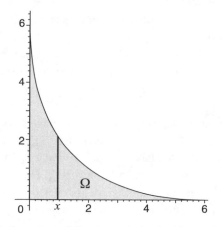

Figure 6.7. At a point on the boundary of Ω, y is maximized subject to the constraint $F(x, y, a) = 0$.

will be parallel precisely when $\partial F/\partial a = 0$. Thus, the Lagrange conditions are

$$F(x, y, a) = 0 \qquad \text{and} \qquad \frac{\partial F}{\partial a} = 0.$$

Since any point (x, y) on the boundary of Ω solves the constrained optimization problem, (x, y, a) must satisfy the Lagrange conditions for some a. This justifies the envelope algorithm, at least for points that can be approached within Ω while holding x fixed. While this result can be extended to other kinds of regions by slightly modifying the argument, it applies as is to our string art example.

6.4   Weighing Means: Arithmetic and Geometric

We have one final example of solving a max/min problem to establish a mathematical result, the well-known Arithmetic Mean - Geometric Mean inequality. Given a set of n nonnegative numbers, the arithmetic mean is just the common average, found by summing the numbers and dividing by n. The geometric mean is the nth root of the product of the numbers, an analogue of the arithmetic mean in which multiplication stands in for addition. Whereas the arithmetic mean is that value that can be substituted for each of the original numbers without changing their sum, the geometric mean can be substituted for each number without changing their product.

Let the n numbers be $x_1, x_2, \ldots, x_n$, where each $x_k \geq 0$. The arithmetic mean is $AM = (x_1 + x_2 + \cdots + x_n)/n$ and geometric mean is $GM = (x_1 x_2 \cdots x_n)^{1/n}$. It is well known that $GM \leq AM$, with equality if and only if all the x's are identical. We will derive this by solving a max/min problem.

It is perhaps not so surprising that a max/min approach is applicable, since the desired result can be stated thus: as a function of n variables, $AM - GM$ has an absolute minimum of 0, assumed whenever the variables are all equal. Indeed, inequalities can generally be seen in a similar light, and so established with max/min problems.

The following derivation, from a paper of Barnier and Martin [10], is cleverer than merely transcribing the Arithmetic Mean - Geometric Mean inequality into a max/min form. It involves the solution of not one max/min problem, but an entire class of them. It is also connected with our previous applications of Lagrange multipliers, and sets the stage for the topic we shall take up next.

Let $f(x_1, x_2, \ldots, x_n) = x_1 x_2 \cdots x_n$, and $g(x_1, x_2, \ldots, x_n) = x_1 + x_2 + \cdots + x_n$. Consider the problem of maximizing f subject to the constraint $g = S$ where S is a constant. Since we assume that all x_k are nonnegative, we can restrict our attention to the case $S \geq 0$.

The constraint set H is an $n - 1$ dimensional hyperplane in $\mathbb{R}^n$. For $n = 3$ it is a triangular region with vertices $(S, 0, 0)$, $(0, S, 0)$, and $(0, 0, S)$, as shown in Fig. 6.8 Because the boundaries lie in the coordinate planes, at each boundary point one of the coordinates is zero. This shows that f vanishes on the boundary of H, and so must take its maximum value in the interior of the triangle, where it assumes only positive values. The same conclusion holds in $\mathbb{R}^n$, where the configuration is analogous.

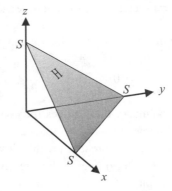

Figure 6.8. The constraint surface H in $\mathbb{R}^3$ has equation $x + y + z = S$.

Suppose the maximum occurs at P. Since it is an interior point, f has a local maximum over H there, and hence the Lagrange multiplier theorem says that ∇f and ∇g are parallel at P. But ∇g is the constant vector $(1, 1, \ldots, 1)$, so the Lagrange condition implies that the components of ∇f must be equal at P. The kth component, $\partial f / \partial x_k$ will be the product of all of the variables other than x_k, and so can be written $\partial f / \partial x_k = f / x_k$. Equality of these components for all k requires equality of the variables x_k. Then the constraint $g = S$ shows that each x_k must equal S/n. We conclude that $(S/n)(1, 1, \ldots, 1)$ is the only solution to the Lagrange conditions, and so must be the location of the maximum value of f subject to $g = S$. At the maximal point, $f = (S/n)^n$.

How does this establish the Arithmetic Mean - Geometric Mean inequality? We argue as follows. Let values of the x's be given, and let S be their sum, so that $AM = S/n$. Then by the max/min problem we just completed, we can see that the maximum possible value of their product is $(S/n)^n = (AM)^n$. This shows that $x_1 x_2 \cdots x_n \leq (AM)^n$, and that equality occurs only if all the x's are equal. Taking the nth root of each side then shows that $GM \leq AM$, as desired.

The appearance of the unspecified parameter S in the derivation is noteworthy. In all our other examples, there was a specific constraint curve C. Solving the Lagrange conditions amounted to selecting a level curve of f that was tangent to C. The preceding argument is different. It concerns an entire family of constraint curves (or more properly in $\mathbb{R}^n$, constraint sets), one for each value of S. Thus we are interested in two families of level sets, those for f and those for g, and points where a member of one family is tangent to a member of the other.

This reveals a kind of symmetry between the objective function and constraint. In fact, reversing the roles of f and g in a constrained optimization problem can lead to the same solution point while providing a completely different perspective. For the Arithmetic Mean - Geometric Mean inequality, we held the sum of the variables fixed and looked for a maximum of the product. With roles reversed we can hold the product fixed and look for a minimum of the sum. Taking both viewpoints into account leads to an idea of *duality*, which is the focus for the next section.

6.5   Isoperimetric Duality

In an isoperimetric problem the goal is to maximize the area of a plane region given the length of the boundary. The classical formulation imposes no constraint on the shape of the boundary, and the maximal area occurs when the region is circular. Variations arise from requiring that the region have a particular shape, or by specifying a part of the boundary. For example, the problem of maximizing the area of a rectangle with a given perimeter, standard fare in calculus courses, is an isoperimetric problem. Another example is the problem of maximizing the area of a rectangular pen that can be enclosed on three sides with a given length of fence and with the fourth side formed by an existing wall or building.

Reversing the roles of constraint and objective leads to problems of minimizing the perimeter of a plane region with a specified area. These are referred to as *dual* isoperimetric problems. As you might expect, if you can solve a dual problem, you can also solve the original (or *primal*) problem. In this section we will generalize from isoperimetric problems to a more general class of constrained optimization problems, observing that duality corresponds to the symmetry between constraint and objective functions mentioned earlier. Part of this development uses the Lagrange multipliers method, following the approach of Segalla and Watson [144].

Before proceeding, we should note that there is another notion of duality in the field of optimization. Every linear programming problem has a corresponding dual problem, and the solution of the first corresponds to a solution of the second. This has been generalized to nonlinear problems. While there are some analogies between the two senses of duality, they are not identical. For our discussion, the primal and dual problems both optimize a function over the same space, whereas in linear programming a dual problem is formulated in a different space than the primal. Thus, the two uses of the term *dual* should be kept distinct.

To illustrate our idea of duality, let us consider again maximizing the product of n nonnegative quantities whose sum is specified. The symbolic formulation of the problem is to maximize $f(x_1, x_2, \ldots, x_n) = x_1 x_2 \cdots x_n$ subject to the constraint $g(x_1, x_2, \ldots, x_n) = x_1 + x_2 + \cdots + x_n = S$. We found that the solution occurs at $AM \cdot (1, 1, \ldots, 1)$ where $AM = S/n$ is the common arithmetic mean for every point satisfying the constraint. The dual problem is to minimize g subject to the constraint that f equals a given constant P. The tangency condition for the dual problem is the same as it was for the primal problem, requiring that ∇f and ∇g be parallel, so we again find that the solution occurs at a point where all the x's are equal. In the dual problem, the constraint says $x_1 x_2 \cdots x_n = P$, so we find the solution point to be $GM \cdot (1, 1, \ldots, 1)$ where $GM = P^{1/n}$ is the common geometric mean for every point satisfying the constraint $f = P$. At the solution point, $g = nGM$.

How do we know that this is a minimum for g? Contrary to the previous case, here the domain is not bounded. Since it includes points of the form $(z, P/z, 1, 1, \ldots, 1)$ for all positive z, both x_1 and x_2 can be arbitrarily large, and similarly for the other x_k. But as any x_k increases without bound, so does g, the sum of the variables. This shows that g goes to infinity in every direction along the constraint surface. On the other hand, g is bounded below by 0. It follows that g must assume an absolute minimum somewhere, and since there is only one solution point for the Lagrange conditions, that must be where the minimum occurs.

This solution to the dual problem leads to a proof of the Arithmetic Mean - Geometric Mean inequality in the same way that the solution to the primal problem did. Let values $x_1^*, x_2^*, \ldots, x_n^*$ be given, and let P be their product. Then $(x_1^*, x_2^*, \ldots, x_n^*)$ satisfies the constraint $f(x_1, x_2, \ldots, x_n) = P$, and so lies in the constraint set for the optimization problem. Thus, nGM, the minimum of g over the constraint set, must be less than or equal to g's value at $(x_1^*, x_2^*, \ldots, x_n^*)$. This gives us

$$nGM \leq x_1^* + x_2^* + \cdots + x_n^*$$

and dividing by n shows that $GM \leq AM$.

Notice that in this argument it is crucial to solve a max/min problem where the constraint value is an unspecified parameter. Then, whatever the product P is for our point $(x_1^*, x_2^*, \ldots, x_n^*)$, we can identify the minimum value for g subject to $f = P$ for that P. The same remarks apply to the earlier derivation of the Arithmetic Mean - Geometric Mean inequality.

Fig. 6.9 illustrates the situation for the two-variable case. The level curves for $f(x, y) = xy$ are hyperbolas with the x and y axis for asymptotes. The level curves for $g(x, y) = x + y$ are lines with slope -1. In the first derivation of the Arithmetic Mean - Geometric Mean inequality, we picked a line, held it fixed, and then looked for the hyperbola tangent to that

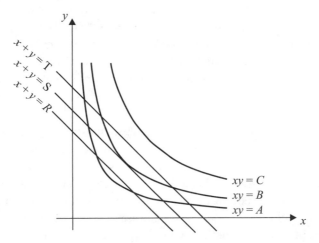

Figure 6.9. Level curves for the sum and product of two variables. At a point of tangency the product is maximized for a particular sum and the sum is minimized for a particular product.

line. For the second derivation we picked a hyperbola and then looked for the line tangent to it. At any point of tangency we get a solution to two dual problems. If $(x_1^*, x_2^*, \ldots, x_n^*)$ is a point where level curves of f and g are tangent, and if $f(x_1^*, x_2^*, \ldots, x_n^*) = P$ and $g(x_1^*, x_2^*, \ldots, x_n^*) = S$, then $(x_1^*, x_2^*, \ldots, x_n^*)$ is the location of both a maximum of f subject to the constraint $g = S$ as well as a minimum of g subject to the constraint $f = P$. This illustrates the idea of duality.

Let us see how duality might be applied in another problem. Suppose we wish to find the maximum area that can be enclosed by a rectangle whose perimeter is 20. With x and y representing the dimensions of the rectangle, the area is $f(x, y) = xy$ and the perimeter is $g(x, y) = 2x + 2y$. The primal problem is to maximize f subject to $g = 20$.

We can as easily solve the more general problem of maximizing f subject to $g = P$, leaving the constant P unspecified. As always, we wish to find points where ∇f and ∇g are parallel. So we compute $\nabla f(x, y) = (y, x)$ and $\nabla g(x, y) = (2, 2)$ and conclude that $x = y$ at the solution point. Combined with the constraint $2x + 2y = P$ this shows that $(P/4, P/4)$ is the only solution, and there $f = P^2/16$. In other words, $f = P^2/16$ is the only level curve of f that is tangent to the curve $g = P$, and the point of tangency is $(P/4, P/4)$. Since we began with $P = 20$, $(x, y) = (5, 5)$ and $f = 25$. Thus, a rectangle with perimeter 20 can enclose an area of at most 25, and to do so it must be a square of side 5.

Solving the general version of the problem allows us to solve instances of its dual as well. Suppose we wish to minimize the perimeter of a rectangle whose area is $A = 36$. Let the unknown minimum perimeter be P, and consider the dual problem: maximize area for that perimeter. As we have seen, the solution occurs at $(x^*, y^*) = (P/4, P/4)$ and the maximal area is $P^2/16$. But we want the area to be 36. So set $P^2/16 = 36$ to determine that $P = 24$ and $(x^*, y^*) = (6, 6)$. This shows that a rectangle of perimeter 24 has maximal area 36. But it also shows that a rectangle of area 36 has minimal perimeter 24, which is what we wanted to know.

This shows how a solution of a constrained optimization problem in a general form can also solve an instance of the dual problem. Each solution to an instance of either problem is also a solution to a corresponding instance of the other.

Here is another way of looking at this situation. Use the functions f and g to define a mapping $\Phi : \mathbb{R}^2 \to \mathbb{R}^2$, with

$$\Phi(x, y) = (P, A) = (2x + 2y, xy)$$

(see Fig. 6.10). The figure shows that Φ maps the closed first quadrant into a shaded region of the P-A plane. One boundary of this region is the positive P axis where $A = 0$. Clearly, Φ maps the x and y axes to this boundary.

The other boundary is the curve $A = P^2/16$. We can see this in several ways. First, we know that when the matrix

$$d\Phi = \begin{bmatrix} \frac{\partial P}{\partial x} & \frac{\partial P}{\partial y} \\ \frac{\partial A}{\partial x} & \frac{\partial A}{\partial y} \end{bmatrix} = \begin{bmatrix} 2 & 2 \\ y & x \end{bmatrix}$$

is singular, (x, y) maps to a boundary point (as we also observed on page 106). This occurs when the determinant of the matrix is zero, and thus when $x = y$. At such points $P = 4x$

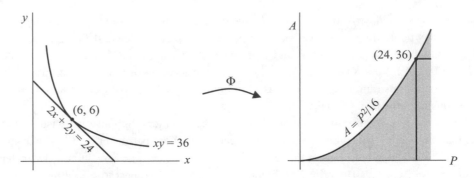

Figure 6.10. Φ takes (x, y) to (P, A). The image of the closed first quadrant in the xy plane is the shaded region in the PA plane.

and $A = x^2$ so $A = P^2/16$. This is one way to see that the curve $A = P^2/16$ is part of the boundary of the image of Φ.

A second approach is to show that in the image of Φ, $0 \le A \le P^2/16$, or in terms of x and y, $0 \le xy \le (x + y)^2/4$. Showing that is a routine algebra exercise. We have already observed that $A = P^2/16$ when $x = y$. Combining both results shows that the shaded region in the figure correctly depicts the image of Φ.

There is a third approach that illuminates the idea of duality. In the figure, the point $(P, A) = (24, 36)$ is marked, along with the line segments in the shaded region where $P = 24$ and $A = 36$. These are the images of the level curves $2x + 2y = 24$ and $xy = 36$ in the domain of Φ. Let us interpret Fig. 6.10 the same way we interpreted Fig. 5.5. Maximizing A while holding P fixed at 24 means moving up the vertical line segment in the shaded region as far as we can. The maximum value corresponds to a point on the boundary of the shaded region, in this case $(24, 36)$. Similarly, minimizing P while holding A fixed at 36 means moving along the horizontal segment as far to the left as possible, again reaching $(24, 36)$. This shows that the point $(24, 36)$ represents both the maximal area for perimeter 24 as well as the minimal perimeter for area 36. In general, every point (P^*, A^*) on the boundary curve $A = P^2/16$ represents the solution to the dual problems of maximizing A subject to $P = P^*$ and minimizing P subject to $A = A^*$. The points (P^*, A^*) are precisely the images $\Phi(x^*, y^*)$ for points (x^*, y^*) where the Lagrange conditions hold.

In this example, we have seen the idea of duality from several different perspectives. Let us proceed to a general formulation of duality. Given suitably defined functions f and g, the problem of maximizing or minimizing f subject to the constraint that g is constant is dual to the problem of maximizing or minimizing g subject to the constraint that f is constant. Any solution of an instance of the first problem is also a solution of an instance of the second. Thus, if $(x_1^*, x_2^*, \ldots, x_n^*)$ is a point at which f assumes an extreme value of f^* subject to the constraint $g = g^*$, then it is also a point at which g assumes an extreme value of g^* subject to the constraint $f = f^*$.

Without considering a formal proof, we can intuitively validate this idea of duality. Suppose that f assumes a maximum value of B subject to $g = S$ at point P, as illustrated in Fig. 6.11. Assuming that the hypotheses of the Lagrange Multiplier Theorem hold, the curves $f = B$ and $g = S$ must be tangent at P. Also at P the gradients of f and g are parallel, and so either point in the same direction or in opposite directions. We consider first the case that they point in opposite directions.

Since f has a constrained maximum of B at P, f can assume values less than B but not greater than B on the constraint curve $g = S$. This shows that in the figure, since the curve $f = A$ intersects the constraint curve, $A < B$. (This is a point that must be argued more carefully in a formal proof.) We conclude that ∇f points upward at P.

Then, because we are assuming that ∇f and ∇g are oppositely directed, ∇g points downward at P, so $R > S$ and $T < S$. Now reverse the roles of objective and constraint functions. Think of the curve $f = B$ as a constraint. At P we know that $g = S$. As depicted in the figure, a level curve where g is greater than S (for example R) does not intersect the constraint curve. So on the constraint, g cannot exceed S. This shows that g attains a constrained maximum at P.

So far, we have argued that if f has a constrained maximum at P and if the gradients of f and g point in opposite directions, then g also has a constrained maximum at P. Using similar reasoning we can argue that when f is a maximum and the gradients point in the same direction, then g has a minimum at P. Analogous results can be justified if f has a minimum rather than a maximum. In either case, when the gradients are opposed, both functions have a constrained maximum or both have a constrained minimum at the point of tangency, while if the gradients point in the same direction when one function has a constrained maximum, the other has a constrained minimum, and vice versa.

The preceding arguments depend on the configuration in Fig. 6.11. We used the fact that the middle f curve intersects the top g curve but not the bottom one. But that need not always happen. For the configuration in Fig. 6.12 the opposite conclusion holds. The middle f curve intersects the bottom g curve but not the top one. This reverses the inferred direction of the gradients of f or g, so that a maximum becomes a minimum or vice versa. Nevertheless, the final conclusions are the same: opposite gradients imply that f and g have the same type of extreme value (both maximums or both minimums) while gradients in the same direction imply that f and g have opposite types of extremes.

It is difficult to succinctly describe all the combinations of curvature, maxima, and minima that can arise. If we consider only extreme values, without distinguishing maxima from

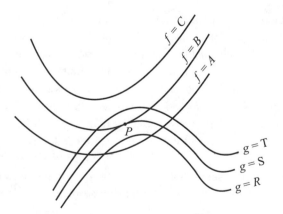

Figure 6.11. Generic configuration for dual constrained optimization problems. Where level curves for f and g are tangent we observe an extremum of f subject to g being constant and vice versa.

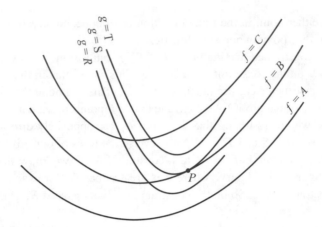

Figure 6.12. Generic configuration where level curves of f and g are all concave up.

minima, the general principle is clear: If (under appropriate hypotheses) f assumes a local extreme value of f_0 at P subject to the constraint $g = g_0$, then g assumes a local extreme value of g_0 at P subject to the constraint $f = f_0$. Thus, these two symmetric optimization problems are duals of one another, and their solutions occur at the same point.

Duality is not a part of the traditional calculus curriculum. When a constrained optimization problem arises, it is considered on its own. There is no hint of the existence of a corresponding dual problem, much less a discussion of what it is and how it is related to the original problem. That is unfortunate because many of the standard problems match up quite naturally in dual pairs. Looking at a problem and its dual together provides additional insight about both.

Even when the interpretation of the dual is not immediately clear, formulating it can provide a new way of thinking about the problem. Consider the milkmaid problem. We wish to find the minimum path length f going from point A to a line L and then to point B. The constraint curve is the line L (where $g = 0$) and the tangent level curve of the path length function is an ellipse.

For the dual problem, we have to restrict consideration to points on the ellipse, and vary the constant value assumed by the constraint function. Making the ellipse the constraint is equivalent to fixing the distance that the milkmaid will walk. Varying the value of g corresponds to translating the line (without altering its slope). So the dual problem is: traveling a fixed total distance from A to C, what is the furthest line parallel to L that can be reached? Here is a version that is similar to the original problem. A milkmaid must drive from A to C with an intermediate stop at a river represented by a line with equation $g(x, y) = D$. She has only enough fuel to travel a total distance of four miles. (Fuel is very expensive!) What is the greatest value of D for which it is possible for her to reach the river?

It is interesting to consider dual problems for all of the familiar questions that are posed in calculus texts. One of my favorites is the ladder problem: how long a ladder can be carried around a corner from a hall of width a into a perpendicular hall of width b? This problem is the focus of Chapter 7. In the meantime, the reader is invited to consider what the dual problem might be.

6.6 A Discontinuous Objective Function

The final topic in this chapter is a max/min problem with an unusual feature — a discontinuous objective function. Actually, such problems are common in practical applications featuring variables that can assume only discrete values. But in textbook examples and exercises the functions that appear are invariably continuous and usually differentiable. The problem we consider here is useful for illustrating that discontinuities can arise naturally, and what that might mean for the optimization techniques covered in calculus.

The Woodsy Shortcut Problem: A hiker wants to go from point A to point B as shown in Fig. 6.13. He can travel along the road by way of point C, or he can take a shortcut through the woods on a heading of angle θ east of north, emerging at some point along the road between C and B. On the road, the hiker can cover 4.5 miles per hour. Going through the woods he can travel only 3 miles per hour. So, while the distance through the woods is shorter, it has to be traveled at a slower pace. If the distance from A to C is 6 miles, and from C to B is 8 miles, how should the hiker go to reach point B as soon as possible?

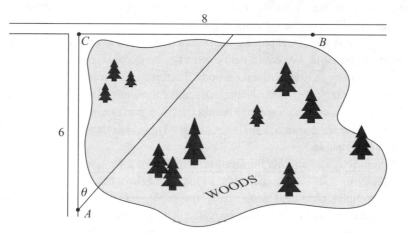

Figure 6.13. The woodsy shortcut problem: can a hiker get from A to B fastest by taking a shortcut through the woods?

As illustrated in the figure, shortcuts may be considered for $0 < \theta \leq \theta_B$, where $\tan \theta_B = 4/3$. The hiker will travel $6 \sec \theta$ miles through the woods and then $8 - 6 \tan \theta$ miles along the road. The first part is traveled at a rate of 3 miles per hour, so the time required is $2 \sec \theta$. For the second part, the hiker goes 4.5 miles per hour with an elapsed time of $(8 - 6 \tan \theta)/4.5$. Thus, the total time for this route will be

$$f(\theta) = 2 \sec \theta + \frac{16}{9} - \frac{4}{3} \tan \theta.$$

For θ between 0 and $\pi/2$, $f'(\theta)$ has one root at $\theta^* = \sin^{-1}(2/3)$, and the second derivative test shows that $f(\theta^*)$ is a local minimum. A computation shows that $f(\theta^*) = (16 + 6\sqrt{5})/9 \approx 3.27$. At the endpoints, we find $f(0) = 34/9 \approx 3.78$ and $f(\theta_B) = 10/3 \approx 3.33$. Thus, it appears that the hiker can reach point B soonest by taking a shortcut through the woods bearing east of north by $\theta^* \approx 41.8°$.

However, this is not correct. Our formula for $f(\theta)$ is not valid when $\theta = 0$. At that heading, the hiker doesn't take a shortcut through the woods at all, traveling all the way along the roads instead. Then the total distance traveled is 14 miles, and at a rate of 4.5 miles per hour that takes $28/9 \approx 3.11$ hours. This shows that the definition for the objective function f should have been

$$f(\theta) = \begin{cases} \frac{28}{9} & \text{if } \theta = 0, \\ 2\sec\theta + \frac{16}{9} - \frac{4}{3}\tan\theta & \text{if } 0 < \theta \le \theta_B, \end{cases}$$

which is not continuous from the right at zero. The optimal path is the one along the road from A to C to B.

This is an instructive example for students. They can discuss the significance of the discontinuity in f, as well as why it arises. For sufficiently small positive values of θ the shortcut route will veer off the road only by a fraction of an inch. This calls into question the assumption, in the original formulation of f, that the hiker covers only three miles per hour on any shortcut route. Students can debate whether or not this is significant in solving the original problem. It leads naturally to a discussion of the assumptions inevitably entailed in modeling a problem mathematically.

Although our formulation expressed the objective function in terms of the angle θ, other approaches are possible. We could define x to be the distance that the hiker must walk along the road after emerging from the woods. Then we can minimize the walking time as a function of x for $0 \le x \le 8$. Mathematically, this formulation is as convenient as the one above. But which one is more useful? Would the hiker prefer to be told *head off at about 42° east*, or to *aim for a spot 5.4 miles west of B*? This is another aspect of modeling that students should consider.

The discontinuity has another implication. Suppose that we ask for the maximum of f, rather than the minimum. In essence, this is looking for the worst case scenario if the hiker makes a poor choice of route. Unexpectedly, this problem has no solution. As θ decreases toward 0, $f(\theta)$ increases toward a limiting value of $34/9$, which is an upper bound for $f(\theta)$ across the entire domain. But f never reaches it because at $\theta = 0$ there is a discontinuous drop to $28/9$. There is no specific value of θ for which f assumes a maximum value. This is a nice illustration of the importance of continuity. A continuous function defined on a closed and bounded interval must assume both a maximum and a minimum, but this conclusion can fail in the absence of continuity.

What a wonderful problem! In one example it offers so many opportunities for discussion: the appearance of a discontinuity in the objective function, its effect on the solution of the problem, issues associated with modeling, and the absence of a definite maximum. If only I could think of a way to work Lagrange multipliers and a dual problem into the analysis, my joy would be complete.

6.7 History, References, and Additional Reading

Optical Property of Ellipses, Rotated Conic Sections.   The idea of basing the optical property of an ellipse on a solution to the milkmaid problem seems to be well known, but I am not aware that this has been previously discussed in connection with Lagrange multipliers. Similar comments apply to finding the axes of a rotated conic. Of course, it may be reasonably objected that using Lagrange multipliers in either context is akin to swatting a fly with a sledge hammer. Having that hammer in hand from the previous chapter, it seemed reasonable to apply it here, and it *does* make for appealingly cute proofs. But it is not at all necessary. For the optical property a simple and understandable proof employing only elementary geometric principles is available in [52], and undoubtedly has appeared elsewhere as well. The same argument is readily found from an internet search. For the rotation angle of a conic, the derivation based on rotating the coordinate axes is not excessively difficult, and can be easily found in calculus textbooks.

I do not know where the milkmaid problem originated. It appears with that name in [144], as well as on numerous internet sites. It also was charmingly posed in a 1917 book on mathematical amusements [41, problem 187]. See Fig. 6.14. Singmaster [149] refers to it as the *shortest route via a wall* problem: Run from A to B touching the wall CD. He gives references dating to 1778.

187.—THE MILKMAID PUZZLE.

In the corner of a field is seen a milkmaid milking a cow, and on the other side of the field is the dairy where the extract has to be deposited. But it has been noticed that the young woman always goes down to the river with her pail before returning to the dairy. Here the suspicious reader will perhaps ask why she pays these visits to the river. I can only reply that it is no business of ours. The alleged milk is entirely for local consumption.

"Where are you going to, my pretty maid?"
"Down to the river, sir," she said.
"I'll *not* choose your dairy, my pretty maid."
"Nobody axed you, sir," she said.

If one had any curiosity in the matter, such an independent spirit would entirely disarm one. So we will pass from the point of commercial morality to the subject of the puzzle.

Draw a line from the milking-stool down to the river and thence to the door of the dairy, which shall indicate the shortest possible route for the milkmaid. That is all. It is quite easy to indicate the exact spot on the bank of the river to which she should direct her steps if she wants as short a walk as possible. Can you find that spot?

HERE is a little pastoral puzzle that the reader may, at first sight, be led into supposing is very profound, involving deep calculations. He may even say that it is quite impossible to give any answer unless we are told something definite as to the distances. And yet it is really quite "childlike and bland."

Figure 6.14. Dudeney's charming version of the milkmaid puzzle.

For more information on ellipses and other conic sections calculus texts may be consulted. In addition, my web-based treatment of Marden's theorem [89] discusses a number of properties of ellipses. It is accessible from the website for this book [87].

String Art.   It is well known that the boundary of a string art design is given by the envelope of a family of curves, and that it can be found with the envelope algorithm. Finding the boundary curve using max/min techniques, and in particular with the Lagrange multiplier method, is something I thought of myself. It seems impossible that no one else has ever had similar ideas, but I do not know of any references.

There is a discussion of envelopes in Chapter 10. Related references to additional reading and historical material are given at the end of that chapter. At the website there is a computer activity that finds and displays envelopes for specified families of curves.

Arithmetic Mean - Geometric Mean Inequality, Duality.   My source for the discussion of the Arithmetic Mean - Geometric Mean inequality is [10], a very interesting paper that shows how several constrained optimization problems can be seen as instances of a single unifying concept. It also suggests several ideas that find more general development in the discussion of duality by Segalla and Watson [144]. The latter is my primary source for duality, although Segalla and Watson do not use that term. What I have called the dual of a max/min problem they refer to as the *flip-side* problem. Their paper also states and proves a theorem detailing how an extremum of the primal gives rise to a corresponding extremum of the dual problem.

It was mentioned earlier that there are two different notions of duality, one histori- cally tied to the isoperimetric problem, the other coming from linear programming. For the isoperimetric problem, the survey by Blåsjö [19] is highly recommended. It is mar- velously written, and won a Lester R. Ford Award. For the other kind of duality, the paper by Strang [155] is a good starting point. It gives an overview of the main ideas outside of the specialized optimization context usually encountered in discussions of duality, and relates them to core concepts from linear algebra.

7

Envelopes and the Ladder Problem

How long a ladder can you carry horizontally around a corner? Or, in the idealized geometry of Fig. 7.1, how long a line segment can be maneuvered around the corner in the *L*-shaped region shown? This familiar problem, which dates to at least 1917, can be found in the max/min sections of many calculus texts and is the subject of numerous web sites. The standard solution begins with a twist, transforming the problem from maximization to minimization. This bit of misdirection no doubt contributes to the appeal of the problem. But it fairly compels the question, *Is there a direct approach?*

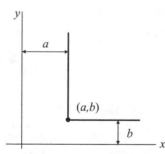

Figure 7.1. Geometry of the ladder problem.

In fact there is one that is beautifully simple and immediately gives new insights about the problem. It depends on an idea from Chapter 6, the envelope of a family of curves. Although this is not a standard topic in calculus courses today, it seems that at one time it was. In any case, defining and finding envelopes is a lovely extension of the current calculus curriculum. It is easily accessible, depending only on concepts from multivariable calculus, and naturally involves graphic images that are highly appealing, like the string art pattern of Chapter 6.

We will explore envelopes in a systematic way in Chapter 10. For now, all we need is the envelope algorithm (page 127). We will apply it to the ladder problem and some extensions.

7.1   A Traditional Approach to the Ladder Problem

The standard solution to the ladder problem begins with a restatement: the goal is shifted from finding the longest ladder that will go around the corner to finding the shortest ladder that will get stuck. We consider line segments through the point (a, b) that reach from the x-axis to the y-axis, as illustrated in Fig. 7.2. As long as the ladder is shorter than all such segments, it can pass around the corner without getting stuck. So, by finding the shortest segment, we determine the maximum length for the ladder.

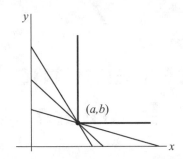

Figure 7.2. Line segments through (a, b) with endpoints on the positive x- and y-axes. The shortest such segment gives the length of the longest ladder that can turn the corner.

There are various approaches to finding the shortest segment, depending on the choice of independent variable. In Fig. 7.3, h, c, d_1 and θ, are all possible choices, as is the slope m of the line through (a, b). For each choice we can express $d_1 + d_2$ (or its square) as a function of the independent variable, and so determine its minimum. Finding it is a routine exercise, although some approaches are algebraically more involved than others.

A particularly elegant solution appears when θ is taken as the independent variable. From Fig. 7.3, $a/d_1 = \cos \theta$ and $b/d_2 = \sin \theta$, leading to

$$d_1 + d_2 = f(\theta) = \frac{a}{\cos \theta} + \frac{b}{\sin \theta}.$$

We wish to find the minimum value of this function for $0 < \theta < \pi/2$. Toward that end, we compute

$$f'(\theta) = \frac{a \sin \theta}{\cos^2 \theta} - \frac{b \cos \theta}{\sin^2 \theta}.$$

This is zero when

$$\tan^3 \theta = b/a, \tag{1}$$

showing that there is exactly one critical point in the domain of f. In fact, $f'(\theta)$ is negative for $\tan^3 \theta < b/a$ and positive for $\tan^3 \theta > b/a$. Therefore, f assumes an absolute minimum at the unique critical point $\theta^* = \tan^{-1}(\sqrt[3]{b/a})$.

To complete the solution, we calculate $f(\theta^*)$. We know that $\tan \theta^* = (b/a)^{1/3}$, and so can find

$$\cos \theta^* = \frac{a^{1/3}}{\sqrt{a^{2/3} + b^{2/3}}} \quad \text{and} \quad \sin \theta^* = \frac{b^{1/3}}{\sqrt{a^{2/3} + b^{2/3}}}.$$

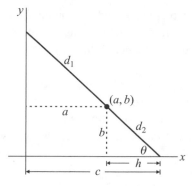

Figure 7.3. The length $d_1 + d_2$ may be minimized as a function of c, h, d_1, θ, or the slope m of the line segment through (a, b).

Thus,

$$
\begin{aligned}
f(\theta^*) &= \frac{a\sqrt{a^{2/3} + b^{2/3}}}{a^{1/3}} + \frac{b\sqrt{a^{2/3} + b^{2/3}}}{b^{1/3}} \\
&= a^{2/3}\sqrt{a^{2/3} + b^{2/3}} + b^{2/3}\sqrt{a^{2/3} + b^{2/3}} \\
&= (a^{2/3} + b^{2/3})^{3/2}.
\end{aligned} \tag{2}
$$

This shows that a line segment can go around the corner if its length is less than or equal to $(a^{2/3} + b^{2/3})^{3/2}$.

The number of choices for an independent variable adds to the richness of the problem (and to its difficulty for students). No matter which variable is chosen, the solution depends on the initial clever transformation, which takes most students by surprise. The original problem clearly asks for a maximum: What is the *longest* ladder that will turn the corner? But we look for a minimum: What is the *shortest* line segment through (a, b) that will touch both axes? Although a little reflection makes it clear that this is a valid approach, on first exposure this reversal can be completely unexpected. "How was I supposed to know to do *that*?" is a natural reaction.

Those who are completely at home with calculus may find some charm in problems with offbeat and unexpected solutions. The inspiration or insight necessary to reach the solution adds to the appeal of the problem. Most students, though, are not charmed. They often view problem-solving as a matter of learning to recognize and apply routine procedures, and prefer to keep creativity and invention to a minimum. The standard solution to the ladder problem offers an opportunity to expand a student's understanding of the nature of mathematical problem-solving, and for this reason it deserves to remain in the curriculum.

Accordingly, what comes next is definitely *not* intended as a "better" approach for the classroom. Rather, it is meant for the old calculus hand, who finds the curriculum familiar and comfortable. Given the reversal inherent in the standard solution, the question is so natural it almost seems to ask itself: *Isn't there a direct approach?* Indeed there is, and it is beautifully simple.

While I do not propose it as a substitute for the standard solution, I do want to remark on one way the direct approach might be relevant to the calculus classroom. Generally, it is good to dissect problems that students find particularly challenging. For the ladder problem, after a solution has been presented, students can be asked to analyze what makes the approach difficult, and whether it offers any lessons for future problems. Once the reversal has been identified and discussed, the possibility of using it in other problems can be explored. Are there other max/min problems in the assignment that are reversible? Students might be challenged to formulate reversed versions of other problems, and to compare the reverse and direct solutions. In discussing these issues the idea of a direct method for the ladder problem might easily arise. If so, it could provide an entrée to interested students, challenging them to learn about the direct approach we will see, or to search for one on their own.

7.2   A Direct Approach to the Ladder Problem

Consider moving a segment around a corner, trying to use as little space as possible. Begin with the segment along one of the outer walls, say with the left end at the origin and the right end at the point $(L, 0)$. Slide the left end up the y-axis, all the while keeping the right end on the x-axis. It seems obvious that this maneuver keeps the line segment as far as possible from the corner point (a, b). If a segment of length L cannot get around the corner this way, then it won't go around no matter what we do.

As you slide the segment along the walls it sweeps out a region Ω, as illustrated in Fig. 7.4. If the corner point (a, b) is outside Ω then all of Ω lies within the hallways, so the sliding segment can go around the corner. On the other hand, if (a, b) falls inside Ω, the segment will *not* go around the corner. It is also clear that increasing the length of the segment expands the region Ω outward from the origin, so that the Ω for one segment is contained within the Ω for any longer segment. We can find the maximum length that will go around the corner by expanding Ω until its boundary curve meets the point (a, b). To do this analytically, we need the equation of the boundary curve.

The region Ω is created by a family of curves, namely the lines containing successive positions of the moving line segment. The boundary curve can be found using the envelope algorithm, which was discussed in Chapter 6. As we shall see, it is given by

$$x^{2/3} + y^{2/3} = L^{2/3}. \tag{3}$$

This curve is an instance of an *astroid*, about which more will be said later. For now, we focus on solving the ladder problem. We know the maximum value of L occurs when (a, b) lies on the boundary curve. So we must have

$$a^{2/3} + b^{2/3} = L^{2/3}.$$

This shows that the longest segment that can go around the corner has length

$$L = (a^{2/3} + b^{2/3})^{3/2}.$$

This elegant solution to the ladder problem depends on (3). How is that derived? To employ the envelope algorithm, we represent the family of curves as an equation $F(x, y, \alpha) = 0$, where each value of the parameter α determines one curve in the family. There are as many choices for α as there were choices of independent variable in the standard solution of the ladder problem. Each member of the family of curves is a line whose x and y intercepts are L units apart. We can identify one of these lines by knowing its x intercept,

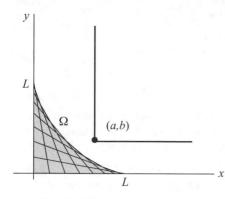

Figure 7.4. Swept out region Ω.

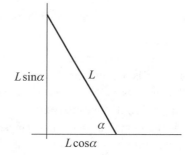
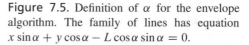

Figure 7.5. Definition of α for the envelope algorithm. The family of lines has equation $x \sin \alpha + y \cos \alpha - L \cos \alpha \sin \alpha = 0$.

y intercept, slope, or the angle between the line and the x-axis. Taking the angle as the parameter makes the envelope algorithm particularly simple.

As Fig. 7.5 shows, the x and y intercepts for the line with angle α are $L \cos \alpha$ and $L \sin \alpha$. This is similar to the example on page 127, and we proceed in the same fashion. Knowing the intercepts tells us that an equation for the line is given by

$$\frac{x}{L \cos \alpha} + \frac{y}{L \sin \alpha} = 1$$

or

$$\frac{x}{\cos \alpha} + \frac{y}{\sin \alpha} - L = 0. \tag{4}$$

If we let $F(x, y, \alpha)$ be the left-hand side of the equation, then $F(x, y, \alpha) = 0$ describes our family of curves.

The next step is to differentiate (4) with respect to α, finding

$$\frac{x \sin \alpha}{\cos^2 \alpha} - \frac{y \cos \alpha}{\sin^2 \alpha} = 0$$

or

$$x \sin^3 \alpha = y \cos^3 \alpha. \tag{5}$$

By combining this equation with (4), we wish to eliminate α. Write (5) in the form

$$\tan^3 \alpha = \frac{y}{x} \tag{6}$$

so

$$\tan \alpha = \frac{y^{1/3}}{x^{1/3}}.$$

This leads to

$$\cos \alpha = \frac{x^{1/3}}{\sqrt{x^{2/3} + y^{2/3}}} \quad \text{and} \quad \sin \alpha = \frac{y^{1/3}}{\sqrt{x^{2/3} + y^{2/3}}}. \tag{7}$$

Substitution in (4) produces

$$x^{2/3} \sqrt{x^{2/3} + y^{2/3}} + y^{2/3} \sqrt{x^{2/3} + y^{2/3}} = L.$$

Simplifying,

$$(x^{2/3} + y^{2/3})\sqrt{x^{2/3} + y^{2/3}} = (x^{2/3} + y^{2/3})^{3/2} = L$$

and so we arrive at (3).

In this derivation there are echos of the standard ladder problem solution discussed earlier, as (6) calls to mind the similar (1). The derivations of (3) and (2) are also similar.

With these observations we can compare the two solutions to the ladder problem. In the direct approach, the optimization part of the problem becomes trivial. It is akin to asking "What is the longest segment that can be contained within the unit interval?" This is nominally a max/min problem, but no analysis is needed to solve it. In the same way, the direct approach to the ladder problem immediately renders the solution transparent once we know the boundary curve for Ω. But it is not fair to claim that this approach eliminates the need for calculus. Rather, the point of application of the calculus is shifted from an optimization question to one of finding a boundary curve. As we have seen, the analysis is similar in both cases.

Two additional points are worth making. First, our derivation of (3) also leads to a nice parametric form. In (7), $\sqrt{x^{2/3} + y^{2/3}}$ appears in each denominator. This is equal to $\sqrt{L^{2/3}} = L^{1/3}$, so substituting and solving for x and y produces

$$\begin{aligned} x(\alpha) &= L\cos^3\alpha \\ y(\alpha) &= L\sin^3\alpha. \end{aligned} \tag{8}$$

In this parameterization, $(x(\alpha), y(\alpha))$ is the point of the envelope that lies on the line corresponding to parameter value α. That is handy because it allows us to associate with each curve in our family the point where it intersects the envelope.

The second point concerns the identity of the boundary curve of Ω. It is an astroid, a prominent specimen in the scrapbook of mathematical curves. Actually, it is the part of the

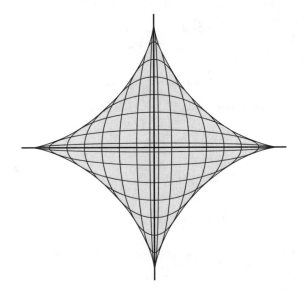

Figure 7.6. The astroid as envelope of a family of ellipses.

astroid that lies in the first quadrant. The full astroid is symmetric in both the x- and y-axes, giving it a star-like appearance and inspiring its name.

To a previous generation of mathematicians who were well acquainted with such terms as *involute*, *evolute*, and *caustic*, an astroid would be familiar indeed. It belongs to a larger class of curves, called *hypocycloids*, because it is the locus of a point on a circle rolling within a larger circle. The astroid occurs when the inner circle has radius exactly one fourth of the outer. The inner circle then executes four full revolutions as it rolls within the outer circle, producing the four symmetric arcs of the astroid.

We obtained the astroid not by rolling a circle in a circle, but as the envelope of the family of lines (4). As shown in Fig. 7.4, the elements of this family create a string art design very similar to the one in Fig. 6.6. Apparently the string art pattern of Fig. 7.4 was known in antiquity, for according to [171] it bears the impressive title the *Trammel of Archimedes*. The astroid can also be obtained as the envelope of a family of ellipses, the lengths of whose axes sum to $2L$. See Fig. 7.6.

We can see that the envelopes in Fig. 7.6 and Fig. 7.4 are the same by verifying that they bound the same regions. We defined Ω as the region swept out by a segment of length L rotated through the first quadrant with its ends on the coordinate axes. Let us consider the part of the region that is produced by just one point on the moving segment. The locus of such a point is an ellipse, the lengths of whose semi-axes sum to L. This fact justifies one of the standard methods for drawing an ellipse (see Sidebar 7.1). Therefore, as we sweep out Ω with our moving line segment, each point of the segment traces an ellipse. The family of such ellipses fills the same region Ω, and so has the same boundary curve. This shows that the family of lines and the family of ellipses have the same envelope curve.

Astroids are worth knowing about for their own sakes. They have pleasing shapes, connect to both a natural string art design and to a broader class of mechanically generated curves, and arise as envelopes for families of lines and families of ellipses. As an added bonus, to those familiar with astroids, the ladder problem becomes utterly transparent. It is clear that the maximum length L occurs when the corresponding astroid passes through the point (a, b). Knowing the equation of an astroid, the extremal condition

$$L^{2/3} = a^{2/3} + b^{2/3}$$

is immediately understood and we can see *why* the maximal L takes the form it does.

7.3 Extending the ladder problem

A variation on the ladder problem is illustrated in Fig. 7.7, where there is a rectangular alcove in the corner where the two hallways meet. The same configuration might occur if there is an obstruction, say a table or a counter, in one hallway near the corner. The problem is to find how long a line segment will go around this corner.

Once again the envelope method provides an immediate solution. Consider the region Ω swept out by a family of lines of fixed length L. If it avoids both (a, b) and (c, d), then the segment can be moved around the corner. As L increases, the envelope (3) expands out from the origin.

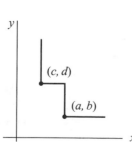

Figure 7.7. Corner with a rectangular alcove.

Sidebar 7.1

Drawing an Ellipse

There are two standard methods for drawing ellipses. The first is based on the definition of an ellipse as the locus of a point from which the distances to two fixed points have a constant sum d (illustrated in Fig. 6.3). To draw the locus, take a string of length d, tack its ends down at the fixed points, and, keeping the string taut around a pencil point, use the pencil to draw an arc.

The second method depends on the characterization of an ellipse as the locus of a point P fixed on a moving line segment that keeps its ends on two perpendicular axes. This is illustrated in Fig. 7.8, where the fixed point divides the segment into pieces of lengths a and b. The figure shows an arbitrary position of the line, making an angle θ with the horizontal axis. If P has coordinates (x, y), then $x/a = \cos\theta$ and $y/b = \sin\theta$. Thus $x^2/a^2 + y^2/b^2 = 1$, which is the equation of an ellipse.

This leads to another way to draw an ellipse. Attach a pencil to a point on a segment, and then slide the segment so that its ends stay on perpendicular axes. Fig. 7.9 is a photograph of a wooden device that is based on this idea.

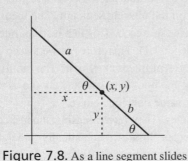

Figure 7.8. As a line segment slides with its ends on the axes, the point (x, y) traces an ellipse.

Figure 7.9. The handle of this mechanical device describes an ellipse.

The maximal feasible L occurs when the envelope first touches one of the corner points. So the maximal value of L is

$$L_{\max} = \min\left\{(a^{2/3} + b^{2/3})^{3/2}, (c^{2/3} + d^{2/3})^{3/2}\right\}.$$

Going further in this direction, we might replace the inside corner with any sort of curve C (see Fig. 7.11). The ladder problem can then be solved by finding the point (x, y) of C for which $f(x, y) = x^{2/3} + y^{2/3}$ is minimized. Unfortunately, this plan is not easy to carry out. Even for the simple case when C is an elliptical arc, the analytic determination of the minimal value of f appears quite formidable, if not impossible. In contrast, the problem is easily solved if C is a polygonal path, because then the minimum value of f will occur at a vertex.

The Couch Problem.   Extending the problem in a different direction, we can make the situation a bit more faithful to the real world by recognizing that a ladder has positive width.

Drawing an Ellipse (cont.)

The handle of the device is attached by screws to guides that run in two perpendicular tracks. As the handle turns, the points of the screws are constrained to run along the tracks, so the section of the handle between the screws behaves like the turning segment I described. Any point on the handle thus traces an ellipse. Fig. 7.10 shows the motion as the handle is turned.

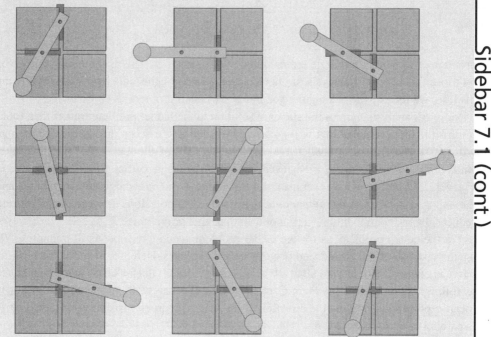

Figure 7.10. Several frames of a movie showing the motion of an ellipse drawing device.

Let us try to maneuver a rectangle rather than a line segment around the corner. If the width of the rectangle is w, what is the greatest length L that permits it to go around the corner?

This version of the problem provides a model for moving bulkier objects than ladders. Trying to push a desk or a couch around a corner in a corridor is naturally idealized to the

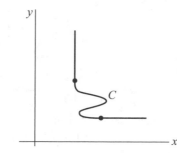

Figure 7.11. Ladder problem with a curve in the corner.

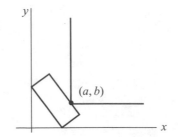

Figure 7.12. This couch is stuck.

problem of moving a rectangle around the corner in Fig. 7.1. This is the motivation given
by Moretti [122] in his analysis of the rectangle version of the ladder problem. Following
his lead, we refer to the rectangular version as the couch problem.

Moretti's analysis mimics the standard solution to the ladder problem: rather than looking for the longest couch that will go around the corner, he seeks the shortest couch that
will get stuck. This occurs when two adjacent corners of the rectangle touch the outer
walls of the corridor and the opposite edge touches the inside corner point, as illustrated in
Fig. 7.12. We will refer to the segment through (a, b) as the *inner* edge of the rectangle and
the opposite segment as the *outer* edge. Using their common slope as a parameter, Moretti
reduces the problem to finding a root of a sixth degree polynomial.

For the couch problem, as for the ladder problem, using envelopes is illuminating. We
again make use of the astroid, and one of its parallel curves. Here, for a curve C, a *parallel*
curve is one whose points are all at a fixed distance from C. Such a curve can be generated
as follows. From each point P on C move a fixed distance w along the normal vector to
locate a point Q. Choosing the direction of the normal vector consistently, the locus of Q
is parallel to C at distance w.

The envelope we need for the couch problem is parallel to the envelope we found for the
ladder problem. This leads to the following geometric interpretation. Move a rectangle of
width w around the corner, sliding the ends of the outer edge along the x- and y-axes, as
in the original ladder problem. Let Ω be the region swept out by the outer edge. If we do
this for a rectangle of maximal length L, then the boundary of Ω must be tangent to the
circle of radius w centered at the point (a, b). See Fig. 7.13. That is, w must be the distance

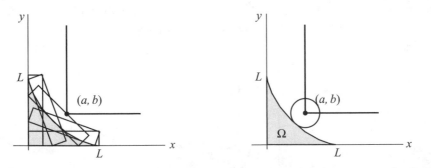

Figure 7.13. Sliding a rectangle of width w and maximum length L around a corner. On the left, it
is shown turning the corner as its outer edge sweeps out Ω. On the right, a circle of radius w centered
at (a, b) is tangent to Ω's boundary. The distance from (a, b) to the boundary equals the width of the
rectangle.

from the corner point (a, b) to the astroid (3). This approach has an appealing simplicity, but requires solving a sixth degree equation, equivalent to the one found by Moretti.

Let us examine the procedure in the figure more carefully. Start with the rectangles's bottom (which is the outer edge) on the x-axis and its left side on the y-axis. Then slide the rectangle so its lower left-hand corner follows the y-axis, and keep the lower right-hand corner on the x-axis. As it slides, the rectangle's bottom edge follows the same trajectory as the segment in the ladder problem, sweeping out the same region Ω. But we want to look at the region swept out by the entire rectangle. Its upper boundary is the envelope of the family of lines corresponding to the inner edge (or initially, the top) of the rectangle. For a couch with length L and width w, these lines can be characterized as follows. Begin with a line in the family for the original ladder problem, whose intersection with the first quadrant has length L. Construct the parallel line at distance w (and in the direction away from the origin). We seek the envelope of the family of all of these parallel lines.

We represent a line in the new family in terms of the same angle α as before (Fig. 7.5). The unit vector parallel to the line is $\mathbf{m} = (-\cos\alpha, \sin\alpha)$, and the normal unit vector (pointing into the first quadrant) is $\mathbf{n} = (\sin\alpha, \cos\alpha)$. These vectors provide a simple way to define a line at a distance d from the origin: begin with the line through the origin parallel to $\mathbf{m}$, and translate by $d\,\mathbf{n}$. Any point on the translated line can be expressed as

$$(x, y) = t\mathbf{m} + d\mathbf{n}.$$

Taking the dot product of both sides of this equation with $\mathbf{n}$ gives

$$\sin\alpha\, x + \cos\alpha\, y = d \tag{9}$$

as the equation of our line.

Lines in the original family are described by (4), which we write

$$\sin\alpha\, x + \cos\alpha\, y = L\sin\alpha\,\cos\alpha.$$

From (9) we see that this line is at a distance $L\sin\alpha\,\cos\alpha$ from the origin. We want the parallel line that is w units further away, and so given by

$$\sin\alpha\, x + \cos\alpha\, y = L\sin\alpha\,\cos\alpha + w.$$

To use the envelope algorithm, let

$$G(x, y, \alpha) = \sin\alpha\, x + \cos\alpha\, y - L\sin\alpha\,\cos\alpha - w.$$

Then, thinking of α as a fixed value, $G(x, y, \alpha) = 0$ defines one line in the family. Similarly, with

$$F(x, y, \alpha) = \sin\alpha\, x + \cos\alpha\, y - L\sin\alpha\,\cos\alpha$$

we obtain the lines in the original family by setting $F(x, y, \alpha) = 0$. It will be convenient in what follows to express these functions in the form

$$F(x, y, \alpha) = \mathbf{n} \cdot (x, y) - L\sin\alpha\,\cos\alpha$$
$$G(x, y, \alpha) = \mathbf{n} \cdot (x, y) - L\sin\alpha\,\cos\alpha - w.$$

Our goal is to find the envelope for the lines defined by G, (hereafter referred to as the envelope for G). To apply the envelope algorithm, we must eliminate α from the equations

$$G(x, y, \alpha) = 0$$

$$\frac{\partial}{\partial \alpha} G(x, y, \alpha) = 0.$$

Rather than do this directly, we can use the fact that we know the envelope for F. Since each line in G's family is parallel to a corresponding line in F's family, and at a uniform distance w, it is not surprising that the envelope of G is parallel to the envelope of F, and at the same distance. That is, if (x, y) is on the envelope of F, then the corresponding point of the envelope of G is w units away in the normal direction.

Let us see that this idea is correct. Given a point (x, y) on the envelope of F, there is a corresponding α such that (x, y, α) is a zero of both F and $\frac{\partial F}{\partial \alpha}$. Then (x, y) is on the line with parameter α, which is tangent to the envelope of F at (x, y). At this point, the line and the envelope share the same normal direction. As observed earlier, the unit normal is given by $\mathbf{n} = (\sin \alpha, \cos \alpha)$. We now consider a new point $(x', y') = (x, y) + w\mathbf{n}$. We wish to show that (x', y') is on the envelope of G.

For that we will use $F(x, y, \alpha) = G(x', y', \alpha)$ and $\frac{\partial F}{\partial \alpha}(x, y, \alpha) = \frac{\partial G}{\partial \alpha}(x', y', \alpha)$. To justify the first of these equations,

$$G(x', y', \alpha) = \mathbf{n} \cdot (x', y') - L \sin \alpha \, \cos \alpha - w$$

$$= \mathbf{n} \cdot [(x, y) + w\mathbf{n}] - L \sin \alpha \, \cos \alpha - w$$

$$= \mathbf{n} \cdot (x, y) + w - L \sin \alpha \, \cos \alpha - w$$

$$= F(x, y, \alpha).$$

To justify the second, recall that F and G differ by a constant and so have the same derivatives. Since $\frac{\partial \mathbf{n}}{\partial \alpha} = -\mathbf{m}$ we get $\frac{\partial G}{\partial \alpha}(x, y, \alpha) = \frac{\partial F}{\partial \alpha}(x, y, \alpha) = -\mathbf{m} \cdot (x, y) - L(\cos^2 \alpha - \sin^2 \alpha)$. Thus

$$\frac{\partial G}{\partial \alpha}(x', y', \alpha) = -\mathbf{m} \cdot [(x, y) + w\mathbf{n}] - L(\cos^2 \alpha - \sin^2 \alpha)$$

$$= -\mathbf{m} \cdot (x, y) - L(\cos^2 \alpha - \sin^2 \alpha)$$

$$= \frac{\partial F}{\partial \alpha}(x, y, \alpha).$$

Together these show that (x, y) is on the envelope of F if and only if (x', y') is on the envelope of G, and that the points (x, y) and (x', y') correspond to the same value of α.

A similar analysis holds for any family of curves F. If G is the family of parallels of the curves in F, all at a fixed distance w, then the envelope for G is the parallel of the envelope of F, at the same distance w. For the couch problem, we can find the envelope of G as a parallel to the known envelope of F.

We have already seen that the envelope of F is parameterized by

$$x = L \cos^3 \alpha$$

$$y = L \sin^3 \alpha.$$

That leads to the parametric description of the envelope of G :

$$x = L \cos^3 \alpha + w \sin \alpha$$
$$y = L \sin^3 \alpha + w \cos \alpha.$$

When L is the solution of the couch problem, (a, b) must lie on the envelope of G. Therefore, we can find L (and also find the critical value of α) by solving the system

$$\begin{aligned} a &= L \cos^3 \alpha + w \sin \alpha \\ b &= L \sin^3 \alpha + w \cos \alpha. \end{aligned} \qquad (10)$$

Eliminating L,

$$a \sin^3 \alpha - b \cos^3 \alpha = w(\sin^2 \alpha - \cos^2 \alpha). \qquad (11)$$

This can transformed into a polynomial by substituting t for $\sin \alpha$ and $\sqrt{1 - t^2}$ for $\cos \alpha$. We obtain

$$at^3 - b(1 - t^2)^{3/2} = w(2t^2 - 1).$$

Isolating the term with the fractional exponent and squaring both sides then leads to

$$(a^2 + b^2)t^6 - 4awt^5 + (4w^2 - 3b^2)t^4 + 2awt^3 + (3b^2 - 4w^2)t^2 + w^2 - b^2 = 0.$$

Though we can solve this numerically (given values for a, b, and w), it is almost certainly impossible to solve it symbolically in general. Thus, we cannot solve (10) algebraically, stymieing our effort to express L in terms of a and b.

Using the parallel curve results, we can now establish the geometric conditions shown in Fig. 7.13. If (a, b) is on the envelope of G, then there is a corresponding point (x, y) on the envelope of F. We know that (x, y) is w units away from (a, b), and that the vector between them is normal to the envelope of F. This shows that the circle centered at (a, b) with radius w is tangent to the envelope of F at (x, y).

Visually, we can now see how to find the maximum value of L. Start with L small enough so that the astroid (3) stays clear of the circle about (a, b) with radius w. Now increase L, expanding the astroid, until the curve just touches the circle. When that happens, the value of L is the solution to the couch problem. See Fig. 7.14.

The visual image of solving the couch problem in this way calls to mind two topics from Chapter 6, duality and Lagrange Multipliers. What we have is a dual constrained optimization problem. The primal problem is to find, for a given L, the point on the astroid

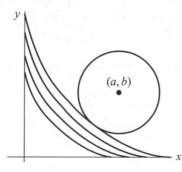

Figure 7.14. Maximizing L geometrically.

$x^{2/3} + y^{2/3} = L^{2/3}$ closest to (a, b). The image for the primal problem is to expand circles centered at (a, b) until one just touches the astroid. For the dual problem we hold the circle fixed and look at level curves for increasing values of the function $f(x, y) = x^{2/3} + y^{2/3}$. We increase the value of f until the corresponding level curve just touches the fixed circle. This corresponds to the dual problem: Find the minimum value of $f(x, y)$ where (x, y) is constrained to lie on the circle of radius w centered at (a, b).

Whichever problem we focus on, the primal or the dual, the solution is the same. For an unknown value of L, we must find a point of tangency between two curves,

$$x^{2/3} + y^{2/3} = L^{2/3} \quad \text{and} \quad (x - a)^2 + (y - b)^2 = w^2.$$

For each curve, we can compute a normal vector as the gradient of the function on the left side of the curve's equation. Requiring them to be parallel leads to the condition

$$x^{1/3}(x - a) = y^{1/3}(y - b).$$

In principle, solving these three equations for x, y, and L would produce the desired solution L to the couch problem. Unfortunately, this seems to lead inevitably to solving a sixth degree equation.

While the envelope method does not provide a symbolic solution to the couch problem, it does give a procedure for generating solvable examples. Here, we will begin with a value of L and produce a triple (a, b, w) so that L is a solution to the (a, b, w) couch problem. To begin, we generate some nice points on the astroid $x^{2/3} + y^{2/3} = L^{2/3}$ using Pythagorean triples. If $r^2 + s^2 = t^2$, then $x = r^3$ and $y = s^3$ give a point on the astroid for $L = t^3$. Thus, using rational Pythagorean triples, we can generate an abundance of rational points on astroids. But the Pythagorean triple does not have to be rational. For example, the irrational triple $(r, s, t) = (3, 4, 5)/\sqrt[3]{5}$ leads to $(27/5, 64/5)$ as a rational point on the astroid curve for $L = 25$.

Combining the equations $x = r^3$, $y = s^3$, $L = t^3$ with the related parameterization

$$x = L \cos^3 \alpha \qquad y = L \sin^3 \alpha$$

of the astroid, we obtain

$$\cos \alpha = \frac{r}{t} \qquad \sin \alpha = \frac{s}{t}.$$

Then the normal vector $\mathbf{n} = (\frac{s}{t}, \frac{r}{t})$, and hence, any value of w leads to the point $(x', y') = (x, y) + w\mathbf{n}$. Take that as the point (a, b). It is w units away from the astroid $x^{2/3} + y^{2/3} = L^{2/3}$, and so lies on the envelope for G. This shows that L solves the (a, b, w) couch problem. We formalize this result in the following theorem.

Theorem.   *For any positive Pythagorean triple (r, s, t) and any positive w, define*

$$a = r^3 + w\frac{s}{t}$$

$$b = s^3 + w\frac{r}{t}$$

$$L = t^3.$$

Then L is the solution to the (a, b, w) couch problem.

For example, with $(r, s, t) = (3, 4, 5)/\sqrt[3]{5}$ and $w = 2$, the equations give $(a, b) = (7, 14)$ and $L = 25$. So for a width of 2, the maximal length of a rectangle that will fit around the corner defined by $(7, 14)$ is 25. In general, if (r', s', t') is a rational Pythagorean triple, and if u^3 is rational, then taking $(r, s, t) = u(r', s', t')$ and rational w produces rational values of a, b, and L, as well as a rational point (x, y) where the astroid meets the circle centered at (a, b) of radius w.

The preceding example, where $L = 25$ solves the $(7, 14, 2)$ couch problem, was given by Moretti. He mentioned that such examples are rare, and asked for conditions on a, b, and w that make the (a, b, w) couch problem exactly solvable. The theorem partially answers this question by providing an infinite family of such triples. It would be nice to know whether every rational (a, b, w) with rational solution L to the couch problem arises in this way. If the critical value of α corresponds to a rational point (x, y) on the astroid (3), then a, b, w, and L are related as in the theorem. But there might be rational (a, b, w) for which the solution to the couch problem is also rational, but which does not correspond to a rational point (x, y).

The envelope approach leads naturally to the theorem, and provides a geometric interpretation of the couch problem solution. Moretti's approach can also lead to an equivalent method for parameterizing triples (a, b, w) with rational solution L. He formulates the problem in terms of a variable m (corresponding to $\cot \alpha$ in our approach) and derives a sixth degree equation in m with coefficients that depend on a, b, and w. Solving for w, one can again parameterize solutions in terms of Pythagorean triples. From this standpoint, the envelope method does not seem to hold any advantage over Moretti's earlier analysis.

7.4 Duality in the Ladder Problem

We saw earlier that the couch problem can be approached using the idea of duality. Now let us answer a question from the previous chapter: what is the dual of the ladder problem?

In the standard approach to the ladder problem we seek the shortest segment through (a, b) that has endpoints on the positive x- and y-axes. We can formulate this as a constrained optimization problem. Introduce new variables u and v, interpreted as x and y intercepts of a line segment in the first quadrant. The objective function, $f(u, v) = \sqrt{u^2 + v^2}$, is the distance between the intercepts. The goal is to minimize f subject to the constraint that the segment must pass through (a, b). Since the line with intercepts u and v line has equation

$$\frac{x}{u} + \frac{y}{v} = 1,$$

that means we must have

$$\frac{a}{u} + \frac{b}{v} = 1.$$

Therefore, letting $g(u, v) = a/u + b/v$, the constraint equation is $g(u, v) = 1$.

This formulation relates two different coordinate planes, as shown in Fig. 7.15. The optimization problem is set in the u-v plane on the right. That is where the functions f and g are defined. The ladder problem interpretation occurs in the x-y plane on the left, where we find the point (a, b) and lines representing positions of the ladder. The connection between the planes is that each point (u, v) on the right corresponds to a line M on the left. In one direction, the coordinates of a point (u, v) become x and y intercepts, specifying a

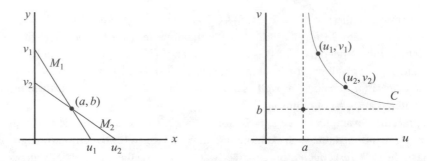

Figure 7.15. Two coordinate planes for minimizing a line segment through (a, b). Each candidate segment M in the x-y plane on the left corresponds to one point in the u-v plane on the right. The intercepts of M are u and v, and we call M the u-v line. The u-v lines through (a, b) on the left correspond to the points (u, v) on the curve C on the right.

line in the x-y plane. We'll call it the u-v line. In the other direction, any line on the left with positive intercepts is the u-v line for some (u, v) in the first quadrant on the right. And the lines through (a, b) on the left correspond to the points (u, v) on the constraint curve C on the right. For these points, $g(u, v) = 1$.

In the standard approach, the primal ladder problem is to minimize $f(u, v)$ subject to $g(u, v) = 1$. For the dual problem, we hold f fixed and look at varying values of the constraint function g. How can this be interpreted? We need to understand what g represents in the x-y context.

Consider a fixed point (u, v), and the corresponding fixed u-v line M in the x-y plane. If $g(u, v) = t$ then $a/u + b/v = t$ so $(a/t)/u + (b/t)/v = 1$. This means that $(a/t, b/t) = (1/t)(a, b)$ lies on M. Here we can think of $1/t$ as a scale factor for moving (a, b) toward or away from the origin in order to reach M. For example, a factor of 2 means we have to double (a, b) to reach M, and $t = 1/2$. A factor of 3 means we have to triple (a, b) to reach M, and $t = 1/3$. In contrast, a factor of $1/2$ means we have to shrink (a, b) by half to reach M, and $t = 2$.

In general terms, the value of g indicates how close the $u - v$ line comes to the point (a, b). Because the reciprocal of t gives the necessary scale factor, $g(u, v)$ is inversely proportional to the distance from the origin to the $u - v$ line along the ray through (a, b). Minimizing g is the same as maximizing the distance.

This gives the following meaning to the dual problem: Look at all the lines whose intercepts u and v are a fixed distance L apart. That means we are looking at points (u, v) for which $f(u, v) = L$. Among all these lines, find the one whose distance from the origin, measured in the direction of (a, b), is a maximum.

The by-now-familiar astroid appears once again as the envelope of a family of lines. Holding f constant with value L (not necessarily the one that solves the ladder problem), we are again considering the family of line segments of length L with ends on the positive x- and y-axes, filling up the region Ω. The point that minimizes g will now be the furthest point you can reach in Ω traveling on the ray from the origin to (a, b). That is, the solution occurs at the intersection of the ray with the envelope (3).

The solution point will be of the form $(a/t, b/t)$ where t is the optimal value of g. Substituting in (3) we find $t = ((a/L)^{2/3} + (b/L)^{2/3})^{3/2}$. This gives us the optimal value of g for the dual problem with an unspecified value of L. To return to the primal problem, we have to choose the value of L that gives us the original constraint value for g, namely $g = 1$. So with $t = ((a/L)^{2/3} + (b/L)^{2/3})^{3/2} = 1$ we again find $L = (a^{2/3} + b^{2/3})^{3/2}$.

It is interesting to compare the various formulations of the ladder problem. The usual approach is to reverse the original problem, so that we seek a minimal line that cannot go around the corner rather than a maximal line that will go around the corner. The envelope approach deals with the problem as stated, finding the longest line that will fit around the corner. A third approach is to take the dual of the reversed version, viewed as an example of constrained optimization. Although all of the approaches are related, each contributes a different understanding of the problem.

7.5   History, References, and Additional Reading

It is not easy to discover when the ladder problem first appeared in calculus texts. Its earliest appearance in Singmaster's extensive chronology [149] of recreational mathematics problems is a 1917 book by Licks [110]. Singmaster notes that Licks' version concerns a stick to be put up a vertical shaft in a ceiling rather than a ladder and two hallways, but the two are mathematically equivalent. Licks gives the standard solution, finding the maximum length stick that gets stuck in terms of the angle the stick makes with the floor. He concludes, "This is a simple way to solve a problem which has proved a stumbling block to many." Whether this implies an earlier provenance in recreational problem solving, or a more mundane history of people actually putting long sticks up vertical shafts, who can say?

American University is fortunate to possess an extensive collection of mathematical textbooks dating to the 18th century. Haphazardly selecting eight calculus textbooks published between 1816 and 1902, I searched without success for mention of the ladder problem in discussions of maxima and minima. Several of the texts had many max/min exercises, including not a few that our students would recognize. Among the 56 max/min exercises in Echols' 1902 text [44], nearly all of today's standard exercises appear, but not the ladder problem. More than half of the books also have a section on envelopes. In the 1862 work of Haddon [62], the astroid arises as an envelope in an example about a ladder sliding down a wall, but not in connection with any max/min problem.

A variant on the couch problem was considered several times in the problems section of the *American Mathematical Monthly* between 1900 and 1940 (see [6] for example). The earliest of these couch problem instances predate the 1917 appearance of the ladder problem cited by Singmaster. During this period, properties of envelopes and the appearance of the astroid seem to have been considered common mathematical knowledge, for the published solutions use them freely without much explanation. It is plausible that the ladder problem should have been as well known as the couch problem, so perhaps the pre-1917 instances of the latter are evidence for equally early appearances of the former.

There does not appear to be a standard version of the *Monthly* problems, where the rectangle that must go around a corner takes many forms: items of furniture, girders, and beams. The different versions are mathematically equivalent, specifying the length and width of the rectangle, and the width of one hallway, and asking for the minimum width of the other hallway that permits the rectangle to turn the corner. In this form, the couch problem can be solved explicitly. In practical terms, either this version or Moretti's version can be applied to determine whether or not a given couch will go around a given corner, though the version with the explicit solution seems aesthetically superior.

In modern times, a variation on the ladder problem has been the subject of ongoing investigation. This is the sofa problem, which seeks the region of greatest area that can go around a corner like the one in the ladder problem. It appears in a volume on unsolved problems in geometry [37], and gave birth to the *Moving a Sofa Constant* [50], defined as the solution when $(a, b) = (1, 1)$. Its exact value is not known, but a lower bound of $2.2195 \ldots$ was obtained in 2003. For more on this open problem, see [167].

The main focus in this chapter has been the use of envelopes to solve the ladder problem. The earliest record I have found for this approach is an anecdote of Cooper [33], who reports meeting a variant of the ladder problem on a physics quiz at Princeton in 1959 and solving it using envelopes. There is also one reference in Singmaster's compilation in which envelopes are used. Fletcher [51] provides five solutions of the ladder problem with no explicit use of calculus, one of which uses envelopes. To avoid calculus, Fletcher depends on geometric properties of envelopes that are little known today. In 1942, Coe [32] used properties of the astroid to solve the ladder problem, which he called the *beam around the corner problem.* However, Coe's solution is quite different from, and more complicated than, the one given here.

The presentation in this chapter made use of properties of envelopes without giving much background on the subject. Also, the idea of parallel curves was mentioned briefly in the discussion of the couch problem. For more on both of these topics a good general reference is [138]. Closer to hand, the topic of envelopes is one focus of Chapter 10 in this book.

For a completely different approach to the ladder and couch problems, using a kinematic concept called the instantaneous center of rotation, see [24].

Most of what appears in this chapter is adapted from [86].

8

Deflection on an Ellipse

"Give me more problems like that!" insisted Mickey. He was an unusual student, a senior history major who signed up for Calculus 3 as an elective, purely because of his interest in mathematics. After class one day he showed me a problem from an earlier assignment like the ones he wanted. He did not care for routine drill problems. He wanted challenging problems, problems that required him to bring together different parts of the course.

The topic for that day's class had been Lagrange multipliers, and under the stimulus of Mickey's request, what I thought would be an interesting application came to mind: to find the maximal deflection between the radial direction and the normal direction at a point of an ellipse. A solution would involve several different ideas he had seen, including normal vectors to curves, angles between vectors, and properties of ellipses. And Lagrange multipliers was an obvious method for maximizing a function over the points of an ellipse. This, I hoped, would be the kind of problem Mickey wanted. It would be a shame to disappoint such a student.

As it turns out, this problem is not a particularly felicitous application of Lagrange multipliers. But it is a good problem, with an interesting answer. It can be attacked from a variety of viewpoints, each of which adds insight. It even has some applied significance (see Sidebar 8.1). Best of all, there is a nice generalization to higher dimensions, with a little bit of a twist.

When I proposed this problem to Mickey, I little suspected how far it would lead. That is the way of Maxministan. All along the main thoroughfares busy travelers pursue their errands, passing side routes and disused turnings with barely a glance. To be sure, some of those paths are mere detours or dead ends. But others are gateways to discovery. They lead to quaint mountain hamlets, serene woodland glades, or breathtaking vistas. In this chapter we will follow the path of Mickey's problem together, thoroughly exploring a little corner of Maxministan.

8.1 The Problem

Here is a careful formulation of the problem. We consider an ellipse, centered at the origin, with semi-major axis a and semi-minor axis b, along the x- and y-axes respectively. The equation of the ellipse is

$$\frac{x^2}{a^2} + \frac{y^2}{b^2} = 1. \tag{1}$$

The vector from the origin to a point of the ellipse defines the radial direction for that point. A vector perpendicular to the tangent line defines the normal direction. We know that at the x- and y-intercepts, the radial and normal directions coincide, so at these points the deflection δ, defined as the angle between normal and radial vectors, is 0. At any other point the deflection will be greater than 0. Our goal is to find the value and location of its maximum. With no loss of generality, we can confine our attention to the part of the ellipse in the first quadrant. In Fig. 8.1 a representative ellipse is shown, with the deflection δ, the normal direction **n,** and the radial direction **r.**

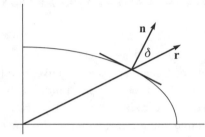

Figure 8.1. Radial and normal directions for an ellipse.

8.2 The Answer

One of the appealing things about Mickey's problem is that the answer has a simple form with a nice geometric interpretation. This does not occur for many problems. In the couch problem (page 148), the solution is a root of a sixth degree equation, and as far as I am aware, it has no explicit symbolic representation, simple or otherwise. In contrast, Mickey's problem has attractive symbolic expressions and simple geometric constructions for the maximal deflection and for the point at which it occurs.

In fact, the maximal deflection has several formulations:

$$\begin{aligned}
\delta_{\max} &= \frac{\pi}{2} - 2\arctan\frac{b}{a} \\
&= \arctan\frac{a}{b} - \arctan\frac{b}{a} \\
&= \arctan\frac{a^2 - b^2}{2ab}.
\end{aligned} \tag{2}$$

It occurs where the ellipse meets the line from the origin to (a, b). So, if you inscribe the ellipse in a rectangle with sides parallel to the axes, and if you draw a line from the center of the ellipse to one of the corners of the rectangle, the line's intersection with the ellipse locates the maximal deflection between the radial and normal directions. See Fig. 8.2. We state this as

Theorem 1. *Let E be an ellipse with center C, semi-major axis a and semi-minor axis b. At any point $P \in E$, let δ be the angle between the normal vector at P and the vector CP. Then the maximum value of δ over E is given by (2). The value is assumed at the points where E intersects the diagonals of the circumscribed rectangle whose sides are parallel to the major and minor axes.*

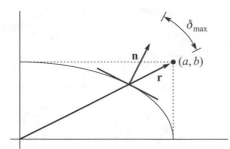

Figure 8.2. Maximal deflection point

The first expression for $\delta_{\max}$ in (2) shows that it increases monotonically with a/b. This is as should be expected: the more eccentric the ellipse, the greater the maximal deflection. Contrariwise, the closer the ellipse is to a circle, the closer the radial and normal directions will be, and so the smaller the maximal deflection will be. Remember this. The monotonicity of the maximal deflection will be important later.

The location of $\delta_{\max}$ given in Theorem 1 suggests a conjecture for n dimensions. In three dimensions the ellipse becomes an ellipsoid (picture a blimp), and the circumscribing rectangle becomes a rectangular box. Over the entire surface of the blimp, where would you find the maximal deflection between the radial and normal vectors? Could it be along a line from the center of the ellipsoid to one of the corners of the box? This is a reasonable guess, based on the two-dimensional case.

As I will explain at the end of the chapter, it is wrong. Happily, there is a beautiful extension to the general case, and you already have enough information to deduce what it is. If you like puzzles and have good geometric intuition, you may want to work out the n-dimensional case before you read further.

8.3 Constructing the Maximal Deflection

Once you know where the maximal deflection occurs, it is easy to construct. The intersection $P = (x^*, y^*)$ of the ellipse and the line through 0 and (a, b) satisfies (1) and the condition $y^*/x^* = b/a$. Thus, $P = (x^*, y^*) = (a, b)/\sqrt{2}$. At P the slope of the radial line is b/a, while from (1) the slope of the tangent line is $-b/a$. This shows that the tangent and radial lines make equal angles α with a horizontal line through P. See Fig. 8.3 where the normal, radial, horizontal, and tangent lines are marked N, R, H, and T. We see at once that $\alpha = \arctan(b/a)$, and also that δ, the angle between R and N, is $\pi/2 - 2\alpha$, verifying the first equation in (2).

This gives a geometric construction for tangent and normal lines at P. Draw the radial line R and the horizontal line H. Duplicate the angle between R and H to construct T. Construct the normal to T to define N.

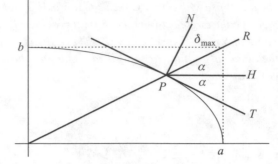

Figure 8.3. Geometric construction of maximal deflection.

There is an alternative construction. At P a line with slope 1 bisects the angle between R and N. Indeed, the slope of R is b/a while the slope of N is a/b, so the angles between them and the x-axis are, respectively, $\arctan(a/b)$ and $\arctan(b/a)$. This gives for δ the expression $\arctan(a/b) - \arctan(b/a)$, and verifies the second equation in (2). It also shows how to construct N and T: draw line R from the origin to (a, b), intersect it with the ellipse to find P, there construct a line with slope 1, reflect R across it to define N, and construct the perpendicular to N to define T.

Both constructions depend only on considering the special point $P = (a, b)/\sqrt{2}$ on the ellipse. We have not yet shown that this is the point where the deflection is maximized. To do so is the object of the original optimization problem. Initially, my intention was that this would be an application of Lagrange multipliers. We will see next how that leads to P. Subsequent sections will show alternative derivations.

8.4   Lagrange Multipliers

To apply Lagrange multipliers, we need to express the problem in terms of an objective function to be maximized and a constraint. Because we consider only points on the ellipse, its equation defines the constraint. Accordingly, let $g(x, y) = \frac{x^2}{a^2} + \frac{y^2}{b^2}$, so the constraint is $g(x, y) = 1$.

Next, let us determine the angle δ between the normal and the radial vectors at a point (x, y) of the ellipse. We may take $\mathbf{r} = (x, y)$ as the radial vector. For the normal vector, take $\mathbf{n} = (x/a^2, y/b^2)$, which is one-half of the gradient of g. Then δ is determined by

$$\cos \delta = \frac{\mathbf{r} \cdot \mathbf{n}}{|\mathbf{r}||\mathbf{n}|}.$$

Since $\mathbf{r} \cdot \mathbf{n} = g(x, y) = 1$, we invert and square to obtain

$$\sec^2 \delta = |\mathbf{r}|^2 |\mathbf{n}|^2 = (x^2 + y^2) \left(\frac{x^2}{a^4} + \frac{y^2}{b^4} \right).$$

We define this to be our objective function, $f(x, y) = (x^2 + y^2)(\frac{x^2}{a^4} + \frac{y^2}{b^4})$. For (x, y) in the first quadrant and on the ellipse, we know that δ is between 0 and $\pi/2$. On this interval $\sec^2 \delta$ is an increasing function. Therefore, δ is maximized where f is.

Geodetic and Geocentric Lattitude

The problem of maximal deflection has an application. It concerns the ellipsoidal model of the Earth, and two ways to define latitude. The following discussion is based on [11, p. 94].

On a sphere, the latitude at a point is the angle between the equatorial plane and the position vector from the center of the sphere. This vector defines the local vertical direction, which is also the normal to the local tangent plane and the (opposite of the) direction of the gravitational force at the point.

However, the Earth is not a sphere. A more accurate model is an ellipsoid, also referred to as an oblate spheroid, with circular cross sections parallel to the equatorial plane and elliptical cross sections of a fixed eccentricity perpendicular to the equatorial plane. For an oblate spheroid, the local vertical direction indicated by a plumb bob does not point toward the center of the Earth. It is normal to the surface of the ellipsoid, which is an idealized level surface with respect to the combination of gravity and the centrifugal acceleration induced by the Earth's rotation.

Because the local vertical direction is easier to measure than the direction to the center of the Earth, it is a convenient reference for defining latitude. The angle between the local vertical direction and the equatorial plane gives the *geodetic* latitude ϕ_d. It is distinct from the more familiar *geocentric* latitude ϕ_c, the angle between the radial direction and the equatorial plane. An exaggerated version of the geometry is shown in Fig. 8.4 for a spheroid more oblate than the Earth.

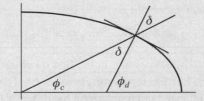

Figure 8.4. Geodetic and geocentric latitude.

As shown in the figure, the deflection δ is the difference $\phi_d - \phi_c$. Theorem 1 thus reveals how far apart the geodetic and geocentric latitudes are at the worst case, and where on the globe that occurs.

The distinction between geodetic and geocentric latitudes is important in determining the locations of celestial objects. Most local observations are made relative to the local vertical, or plumb bob direction. To reconcile observations from different points on the globe, or to register them in a global geospatial model, we have to take into account the deviation between the radial and plumb bob directions.

Sidebar 8.1

Our problem now is to maximize f subject to the constraint $g = 1$. The solution occurs at a point where ∇f and ∇g are parallel. Since vectors (u, v) and (p, q) are parallel just when $uq = pv$, this leads to

$$\frac{\partial f}{\partial x} \frac{\partial g}{\partial y} = \frac{\partial f}{\partial y} \frac{\partial g}{\partial x}.$$

From this it is straightforward (if slightly complicated) to derive

$$\frac{y}{x} = \pm\frac{b}{a}. \tag{3}$$

To do this, first compute the partial derivatives

$$\frac{\partial f}{\partial x} = \frac{2x[2b^4x^2 + (a^4 + b^4)y^2]}{a^4b^4}$$
$$\frac{\partial f}{\partial y} = \frac{2y[2a^4y^2 + (a^4 + b^4)x^2]}{a^4b^4}$$
$$\frac{\partial g}{\partial x} = \frac{2x}{a^2}$$
$$\frac{\partial g}{\partial y} = \frac{2y}{b^2}.$$

These give

$$\frac{\partial f}{\partial x}\frac{\partial g}{\partial y} = \frac{4xy}{a^4b^4} \cdot \frac{2b^4x^2 + (a^4 + b^4)y^2}{b^2}$$

and

$$\frac{\partial f}{\partial y}\frac{\partial g}{\partial x} = \frac{4xy}{a^4b^4} \cdot \frac{2a^4y^2 + (a^4 + b^4)x^2}{a^2},$$

which are equal if and only if

$$2a^2b^4x^2 + (a^4 + b^4)a^2y^2 = 2a^4b^2y^2 + (a^4 + b^4)b^2x^2.$$

After one more rearrangement we obtain

$$a^2y^2(a^4 - 2a^2b^2 + b^4) = b^2x^2(a^4 - 2a^2b^2 + b^4),$$

from which (3) is apparent. This shows that in the first quadrant, the solution to our optimization problem lies on the line joining the origin to (a, b). The algebra in the preceding derivation it is not too complicated to complete by hand, but it is sufficiently involved to make us want to consider alternatives.

8.5 Direct Parameterization

The standard parameterization of the ellipse, namely,

$$(x, y) = (a\cos t, b\sin t), \tag{4}$$

provides an alternative to Lagrange multipliers. Substitution in the objective function leads to

$$\sec^2\delta(t) = f(a\cos t, b\sin t) = \cos^4 t + \cos^2 t \sin^2 t \left(\frac{a^2}{b^2} + \frac{b^2}{a^2}\right) + \sin^4 t,$$

expressing f as a function of a single variable.

This can be reduced to a much simpler expression for $\tan\delta(t)$. Apply the squared Pythagorean identity $(\cos^2 t + \sin^2 t)^2 = 1$, leading to

$$\sec^2\delta(t) = 1 + \left(\frac{a^2}{b^2} + \frac{b^2}{a^2} - 2\right)\cos^2 t \sin^2 t$$

and hence

$$\tan^2 \delta(t) = \left(\frac{a^2}{b^2} + \frac{b^2}{a^2} - 2 \right) \cos^2 t \sin^2 t.$$

Next, the coefficient of $\cos^2 t \sin^2 t$ can be combined into a single fraction. We have

$$\frac{a^2}{b^2} + \frac{b^2}{a^2} - 2 = \frac{a^4 + b^4 - 2a^2 b^2}{a^2 b^2} = \left(\frac{a^2 - b^2}{ab} \right)^2.$$

Finally, take square roots and use the double angle identity for the sine, obtaining

$$\tan \delta(t) = \sin(2t) \frac{a^2 - b^2}{2ab}.$$

To find the maximum value of δ, it suffices to find the maximum of $\tan \delta(t)$, for $0 \leq t \leq \pi/2$. This we can do by inspection. Over the interval, $\sin 2t$ has a unique maximum of 1 at $t = \pi/4$ so

$$\delta_{max} = \arctan \left(\frac{a^2 - b^2}{2ab} \right),$$

one of the expressions in (2). Where does this occur on the ellipse? Substituting $t = \pi/4$ in the parameterization (4) leads to $(x, y) = (a, b)/\sqrt{2}$. This is the same result we found earlier.

It is a common misconception that the variable t in the standard parameterization of an ellipse is equal to the polar angle θ. This is only true for a circle, when $a = b$. Otherwise, at $(x, y) = (a \cos t, b \sin t)$ we have $\tan \theta = y/x = (b \sin t)/(a \cos t) = (b/a) \tan t$ so t and θ can agree only at multiples of $\pi/2$. The difference α between t and θ is defined at each point of the ellipse, vanishing only on the axes. Maximizing it is similar to maximizing δ, and has another appealing solution. This is left as an exercise for the interested reader.

That the maximal deflection occurs at the midpoint of the parameter domain is striking, and suggests searching for a geometric interpretation. In fact, there is a kind of symmetry that makes the solution point at $\pi/4$ very natural. It will be discussed after considering another method for maximizing δ.

8.6   Using Slopes

While the direct parameterization worked out nicely, there is another approach worth considering. It expresses everything in terms of slopes, and uses only ideas from the first calculus course. Consider a point (x, y) on the ellipse, and observe that the slope of the radial line there is $m = y/x$. Next compute m_N, the slope of the normal line, as follows: by implicit differentiation, the slope of the tangent line to the ellipse is $-\frac{b^2 x}{a^2 y}$, so $m_N = (a^2/b^2)m$.

Using their slopes, we can compute the tangent of the angle between the lines from

$$\tan(\delta) = \frac{m_N - m}{1 + m_N m},$$

which leads to

$$\tan(\delta) = \frac{m(a^2 - b^2)}{b^2 + a^2 m^2}.$$

The maximum value of δ occurs at the same point as the minimum of $f(m) = (a^2 - b^2) \cot(\delta) = a^2 m + b^2/m$. Differentiation gives us $f'(m) = a^2 - b^2/m^2$, which vanishes when $m = \pm b/a$. This is, of course, consistent with what we found before.

8.7 Symmetry

We come now to a kind of symmetry that illuminates the location of the maximal deflection point. Using the standard parameterization $\mathbf{r} = (a\cos t, b\sin t)$, we may take $\mathbf{r}$ itself as a radial vector at any point on the ellipse. Differentiating with respect to t produces the tangent vector $(-a\sin t, b\cos t)$, from which we obtain the outward normal vector $\mathbf{n} = (b\cos t, a\sin t)$.

These give rise to a revealing geometric interpretation. Consider Fig. 8.5, which depicts circles of radius b and a centered at the origin. Rectangle $PQRS$ has vertices

$$
\begin{aligned}
P &= (b\cos t, b\sin t)\\
R &= (a\cos t, a\sin t)\\
Q &= (a\cos t, b\sin t)\\
S &= (b\cos t, a\sin t).
\end{aligned}
\tag{5}
$$

As t varies, P and R trace out the circles, and the vectors OP and OR are parallel with polar angle t. Since OQ is the vector $\mathbf{r}$, Q traces our ellipse. S has two interpretations. On the one hand, S traces a second ellipse, with horizontal semi-minor axis b and vertical semi-major axis a. It is the reflection of the first ellipse in the line $y = x$. On the other hand, OS is the vector $\mathbf{n}$, normal to the original ellipse at Q. Thus, the deflection angle δ between the radial and normal vectors appears as $\angle SOQ$. As the parameter t varies from 0 to $\pi/2$, OR sweeps around the outer circle at a constant rate, while $\square PQRS$ evolves continuously from a horizontal segment, through a progression of rectangles, to a vertical segment. In the process, $\angle SOQ$ portrays the variation of δ.

This is where symmetry makes its entrance. Fig. 8.6 shows $\square PQRS$ and $\square P'Q'R'S'$ corresponding to parameter values t and $t' = \pi/2 - t$. Since $t + t' = \pi/2$, the rectangles are mirror images in the line $y = x$. Reflection in the line preserves the identities of the P and R points, while interchanging Q and S. But $\angle SOQ$ and $\angle S'OQ'$ are the same, so δ assumes equal values for t and t'. Thus, $\delta(t)$ is symmetric with respect to $t = \pi/4$.

This is not sufficient to tell us that the maximum value of δ occurs at $t = \pi/4$. But if it occurs elsewhere, there must be two solution points that are symmetric about $\pi/4$. So if there is a unique maximum, symmetry shows it occurs at $t = \pi/4$.

The geometry of Fig. 8.6 also provides an alternate derivation of an expression we found for $\tan\delta(t)$. Looking at Fig. 8.5, we know that $\delta = \angle SOQ$. We can obtain this as the

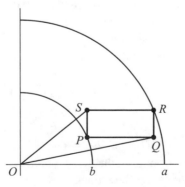

Figure 8.5. Parametric rectangle.

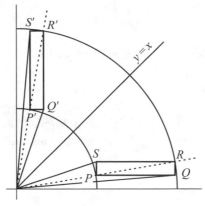

Figure 8.6. Symmetric rectangles.

difference between the angles that OS and OQ make with the x axis. Call these θ_S and θ_Q, respectively. Thus $\tan \delta = \tan(\theta_S - \theta_Q)$, or

$$\tan \delta = \frac{\tan \theta_S - \tan \theta_Q}{1 + \tan \theta_S \, \tan \theta_Q}. \tag{6}$$

From (5), $\tan(\theta_S) = (a \sin t)/(b \cos t)$ and $\tan(\theta_Q) = (b \sin t)/(a \cos t)$. Substitution in (6) produces

$$\tan \delta = \sin(2t)\frac{a^2 - b^2}{2ab}.$$

This validates the intuition that δ increases from 0 to a unique maximum and then decreases back to 0 as t goes from 0 to $\pi/2$. Now the symmetry argument shows that the maximum must occur at $t = \pi/4$.

8.8   Generalization to $\mathbb{R}^n$

There is a generalization of the problem of maximal deflection to n-dimensions. We consider the ellipsoid with semi-axes $a_1, a_2, \cdots a_n$, whose equation is

$$\frac{x_1^2}{a_1^2} + \frac{x_2^2}{a_2^2} + \cdots + \frac{x_n^2}{a_n^2} = 1, \tag{7}$$

and we wish to locate the point where the angle between the normal and radial vectors is greatest. It is tempting to conjecture that the solution lies at the intersection A of the ellipsoid and the segment joining the origin to $(a_1, a_2, \cdots, a_n)$.

Unfortunately, none of our solution methods lends itself to a solution in n dimensions. We saw one approach in which everything was formulated in terms of slope. This does not extend in an obvious way to higher dimensions. Direct parameterization is a possibility, using for example n-dimensional spherical coordinates, but the algebra quickly gets out of hand. Lagrange multipliers is the method that extends most easily to n dimensions, at least as far as formulating the necessary condition for an extreme point. Unfortunately, I have been unable to solve the Lagrange equations. However, it is easy enough to check that the coordinates of A do *not* satisfy the Lagrange conditions. So at least we can determine that the obvious conjecture is false.

James E. White

James White (1946–2004) was a mathematician who was keenly interested in the use of computer technology for mathematical exploration and discovery. He developed the Mathwright family of software applications nearly single-handedly, breaking new ground in educational software. The Mathwright platform empowered math teachers to create interactive computer environments, bringing mathematical objects and their relationships to life. Many of the animations presented at the website for this book [87] were created using Mathwright.

White

White was a topologist by training, completing his Ph.D. at Yale University under William Massey in 1972. He held permanent and visiting faculty positions at many institutions, including the University of California San Diego, Carleton College, Bates College, Kenyon College, Spelman College, California State University Monterey Bay, and Stetson University. His non-academic experience included work at the Jet Propulsion Laboratory. Major work on the Mathwright project occurred during his tenure at the Institute for Academic Technology at the University of North Carolina Chapel Hill.

For more about White and Mathwright, see [83] and [84]. Sadly, Mathwright is no longer being maintained, and will likely become obsolete in the near future.

In spite of these apparent difficulties, the n-dimensional case is simple once you have the right geometric insight. For me, it occurred when a colleague produced an illustration of the geometry in three dimensions. Using Mathwright, the software he created, James White implemented a routine to color-code the surface of a three-dimensional ellipsoid according to the size of the deflection δ, and to view the result from any angle. One sample of this coloring appears in Fig. 8.7. The points that are brightest correspond to the largest values of δ. At the ends of the axes of the ellipsoid, where δ is 0, the shading is darkest.

As suggested by the figure, δ achieves its maximum values in a plane cross section of the ellipsoid, corresponding to the greatest and least semi-axes. This is a consequence of the monotonicity of $\delta_{\max}$ with eccentricity, and leads to the following theorem.

Theorem 2. *On the ellipsoid*

$$\frac{x_1^2}{a_1^2} + \frac{x_2^2}{a_2^2} + \cdots + \frac{x_n^2}{a_n^2} = 1,$$

the maximum value of δ can be found as follows: Consider the plane cross section of the

Figure 8.7. Shaded ellipsoid.

ellipsoid determined by the axes corresponding to the greatest and least values of a_i. The cross section is an ellipse. The maximum value of δ on the ellipse is also the maximum of δ over the entire ellipsoid.

Proof. Let a^* be the maximum of the a_i, and let b^* be their minimum. By renumbering if necessary, we may assume that $a^* = a_1$ and $b^* = a_n$. In the plane determined by the x_1- and x_n-axes, the ellipsoid's cross section is an ellipse E^* with semimajor axis a^* and semi-minor axis b^*. By Theorem 1, the maximum value of δ on E^* is $\delta^* = \pi/2 - 2\arctan(b^*/a^*)$. We wish to show that this is the maximal deflection over the entire ellipsoid.

Consider an arbitrary point P on the ellipsoid. At P the radial and normal vectors determine a plane that intersects the ellipsoid in an ellipse through P centered at the origin. Call the ellipse E; let a be its semi-major axis and b its semi-minor axis. Invoking Theorem 1 again, the maximal deflection over E is $\pi/2 - 2\arctan(b/a)$, and this must be greater than or equal to δ_P, the deflection at P.

To complete the proof, compare E^* with E. No point of the ellipsoid is further from the center than a^* nor closer to the center than b^*. Therefore, $a \le a^*$ and $b \ge b^*$. This shows that $a/b \le a^*/b^*$, so the maximum value of δ on E is less than or equal to the maximum on E^*. Together, these considerations imply that

$$\delta_P \le \max_E \delta \le \max_{E^*} \delta = \delta^*.$$

Therefore, δ^* is the maximum value of δ over the entire ellipsoid. ∎

This brings to an end my account of the surprisingly fertile question: Where on an ellipse is the angle between the radial and normal vectors the greatest? We have seen several methods of solution, an unexpected appearance of symmetry, extensions to higher dimensions, and even an application. There remain other facets of this problem left to explore. For example, there is a related problem: parameterizing an ellipse in the form $(x, y) = (a\cos t, b\sin t)$, find the maximum difference between the parameter t and the polar angle θ. Similarly, we might ask when the area of rectangle $PQRS$ (Fig. 8.5) is maximal.

For another variation, what happens when the origin is at a focus of the ellipse, instead of at its center? In this case, the deflection angle has a connection with astrodynamics. We may imagine the ellipse as the orbit of a planet or a satellite. In astrodynamics, the *flight-path angle β* is defined as the angle between a body's position and velocity vectors. Since the velocity vector is tangent to the ellipse, $\beta = \pi/2 \pm \delta$. Thus the maximum deflection corresponds to a maximum or minimum of the flight-path angle.

Alas we cannot pursue these questions now. Let them remain for additional exploration by the adventurous reader. It is time for our tour to go on to another realm, the Calculusian Republic.

8.9   History, References, and Additional Reading

I am not aware of other investigations of what I have called Mickey's problem, though it would not surprise me to learn that it has been studied before. An account of this material appeared in my paper [85]. Ellipses and the other conic sections have countless interesting properties and are featured in incalculably many problems. Here are a few examples that are closely related to the ideas we have considered in this chapter.

An article by Klamkin and McLenaghan [97] concerns normal lines to an ellipse, and their maximal distance from the center of the ellipse. When the normal line and the radial line are close together, the normal line will pass close to the origin. Thus, a large distance from a normal line to the origin should go with a large value of the deflection δ. They are not maximized at the same point, however. In terms of the polar angle θ, δ is a maximum when $\tan \theta = b/a$, whereas the distance from the normal line to the origin is maximized when $\tan^2 \theta = b/a$.

In our consideration of Mickey's problem, we generalized from an ellipse in two dimensions to an ellipsoid in n dimensions. Sekino [145] generalizes in a different way, to plane curves with n foci. Given n foci, he defines a curve as the set of points for which the sum of the distances to the foci is constant. It would be interesting to consider Mickey's problem in this context.

We used the existence of symmetry in Mickey's problem to help explain the nature of the solution. For a related discussion of symmetry in optimization problems, see [164].

Part III

The Calculusian Republic

CALCULUSIA is vast. It is so large, and has so many regions, landforms, cultures, and climates, that visitors can return again and again without seeing all of the most popular attractions. Not all of these attractions are completely original. Some are highly derivative, and recognizing this should be integral to the visitor's impression of the realm. Nevertheless, it would be unwise to try to digest all that the republic has to offer at a single go. Savvy travelers to Calculusia are wise to know their limits.

Those embarking on a first encounter with Calculusia readily find literature to guide them on their journeys. There are scores of encyclopedic tour books offering comprehensive guidance for the traveler. These are tomes, to be sure: volumes so big that to get them from one place to another you practically have to tow 'em. Among their pages one can continue to discover new (or forgotten) wonders, even after years of study. Rare, indeed, is the mathematical traveler for whom calculus texts offer nothing new.

Books so big you have to tow 'em.

Does this mean that no additional tour guides are needed? If the most popular attractions are already too much to take in, what is the point of exploring other less important sites? These are fair questions, with answers that will vary from one traveler to another. For myself, there is an added charm in pursuing the road less traveled. Going off in a different direction, leaving behind the tumult and the crowds, one gets a sense of personal discovery, the feeling of seeing something that few others have seen.

In so vast a land as Calculusia, the opportunities for exploration and discovery are likewise vast. Surrounding every major destination there are countless lesser known districts to investigate. No, it is not possible to explore all of them, or even a large fraction of them. But the fact that they abound in such richness contributes to the overall mystique of this land of Calculusia. It would be a shame to focus solely on the mainstream, never suspecting the breadth and depth of these hidden riches.

In this final part of the book, we will explore a few out-of-the-way parts of the Calculusian Republic. We begin on the border with analysis, where first-time visitors invariably take a pre-Calculusian tour of orientation. There we will investigate a function called *glog*, a cousin of the natural logarithm that is definitely outside the mainstream. Next, it is on to geometric aspects of curves, concerning envelopes and asymptotes. In a third excursion we will see how derivatives can be developed without limits. Then it is back to the heart of the republic, to witness two miracles of calculus — a fitting finale for our tour of Calculusia.

9

A Generalized Logarithm for Exponential-Linear Equations

How do you solve the equation

$$1.6^x = 5054.4 - 122.35x? \tag{1}$$

At first glance, the appearance of x in the exponent suggests taking a logarithm of both sides, but that approach reaches a dead end almost immediately. The standard tools of precalculus, algebraic manipulations, are unable to isolate x. The difficulty stems from the appearance of exponential and linear expressions in the same equation.

Of course, we can find an approximate solution to the equation using a numerical approach, such as Newton's method or systematic trial and error. Many graphing calculators have built-in root finders that do this automatically, at the touch of a button or two. Using one it is easy to find that the solution to (1) is approximately 17.0152. But this does not have the same appeal as *solving* the equation. Wouldn't it be better to derive a solution analytically using an algebraic process?

If the equation cannot be solved using standard tools, maybe we should expand the toolbox. In this chapter we look at a new function, which I call *glog* (and pronounce "gee-log"), that can solve equations like (1). In particular, glog can be used to solve any equation of the form

$$a^x = mx + b$$

where a, b and m are constants. We will refer to these as *exponential-linear* equations.

9.1 Expanding the Tool Box

Inventing a new function to solve a new type of equation is a venerable strategy in elementary analysis. For example, the square root operation can be regarded as a function introduced to solve quadratic equations. How does that work? By its definition, the square root function produces the unique nonnegative root to the quadratic equation $x^2 = b$. This can be understood by using the notation and terminology of functions. Let $s(x) = x^2$.

175

Then, since $x^2 = b$ can also be expressed as $s(x) = b$, solving it is the same as inverting the s function. Indeed, when we say for nonnegative a and b, $a = \sqrt{b}$ if and only if $b^2 = a$, we are defining $\sqrt{x}$ as the inverse function of $s(x)$. This provides an algebraic solution of any equation $x^2 = b$, although the validity of that solution reduces to a tautology.

However, it turns out that once the square root function is defined, we can solve not just $x^2 = b$, but *any* quadratic equation. Moreover, the square root function has convenient algebraic properties that contribute to its utility in simplifying and solving equations.

Similarly, the definition of $\sqrt[n]{x}$ for positive integers n can be understood in two ways, either as an inverse of the function x^n, or as a special function introduced to permit solution of $x^n = b$. We know that nth roots share the same algebraic properties as $\sqrt{x}$, although it is not true, in general, that these functions can be used to solve arbitrary polynomials. We saw this in Chapter 4. Restricting our attention to solving cubics, the curly root function $\{\overline{x}$ (page 26) is better than $\sqrt[3]{x}$ as an analog of $\sqrt{x}$. Like both $\sqrt[3]{x}$ and $\sqrt{x}$, $\{\overline{x}$ is defined as an inverse function. Specifically, it is the inverse of $f(x) = x^3/(1-x)$. But we can also view it as a special function designed for solving equations of the form $x^3 + bx - b = 0$ for any b. That is, x is a solution to this equation just when $f(x) = b$. And once the curly root function is defined, it can be used to solve all cubic equations.

Another example (which some would no doubt consider less radical) is the natural logarithm, $\ln x$. Not to belabor the point, it is defined as the inverse of the exponential function e^x, and so permits solution of any equation of the form $e^x = b$. But it can do more than that. Combining logarithms and algebraic manipulation permits us to solve any exponential equation $Ac^x = b$.

In each example we have the same general framework. We wish to solve a certain class of equations. We notice that equations in a subclass can be solved by inverting a particular function. Then it turns out that the inverse function enables us to solve all the equations we originally had in mind. Why not try to use this framework for exponential-linear equations?

We can and will, discovering naturally the function I call *glog*. It can be used to solve exponential-linear equations but that is by no means all it is good for. With glog we can write a closed-form expression for the iterated exponential $(x^{x^{x^{\cdot^{\cdot^{\cdot}}}}})$ and solve $x + y = x^y$ for y. The glog function is related to another special function, called the Lambert W function in [22] and [34], whose study dates to work of Lambert in 1758 and of Euler in 1777. Interesting questions about glog arise at every turn, from symbolic integration, to inequalities and estimation, to numerical computation. Many of these points will be elaborated in this chapter. As a first step, we proceed to the definition of glog.

9.2 Defining glog

Let us focus on an especially simple form for an exponential-linear equation:

$$e^x = cx. \tag{2}$$

Solving equations of this form amounts to inverting the function e^x/x : given a particular c, if we can find an x for which $e^x/x = c$, then that x is a solution to (2). Thus, in analogy with the examples given earlier, we are prompted to adopt the following definition of glog:

$$y = \mathrm{glog}(x) \ \text{ iff } \ x = e^y/y \ \text{ (or, iff } e^y = xy).$$

Who Needs glog?

How do we know that a new function really is needed to solve exponential-linear equations? Our inability to find a solution by conventional means is no proof that none exists.

To make this point clearer, let us consider an analogous situation. Suppose we wish to solve equations of the form $x^3 + bx - b = 0$, and suppose we had never heard of Cardano's solution of the cubic. Using algebraic methods, we could labor long and hard without finding a solution. This might tempt us to conclude that solving our cubic using elementary algebra is impossible, and so to define the curly root function as the inverse of $x^3/(1-x)$ (page 26). By this definition, $\{\overline{b}$ solves our equation $x^3 + bx - b = 0$.

But as it happens we *do* know about Cardano's method. Applying it to $x^3 + bx - b$ shows that

$$\{\overline{b} = \sqrt[3]{\frac{b}{2} + \sqrt{\frac{b^2}{4} + \frac{b^3}{27}}} + \sqrt[3]{\frac{b}{2} - \sqrt{\frac{b^2}{4} + \frac{b^3}{27}}},$$

at least for $b \geq -27/4$. Thus, in this case we know that introducing a new function is not necessary — it is already available as a combination of known functions.

Might the same thing be true for exponential linear-equations? Is it possible that an as yet unknown method might exist for solving these equations algebraically? This is a tricky question, akin to asking whether polynomials can be solved in terms of radicals. And it has just recently been answered, in the negative: exponential-linear equations cannot be solved with algebraic methods using just the elementary functions. A proof appears in [27].

Sidebar 9.1

This definition does not give much leverage in evaluating glog. For example, suppose we wish to find glog(2). While the definition identifies glog(2) as the solution to $e^t = 2t$, that doesn't help us find it. But this is getting a bit ahead of the game. The point is to study the properties of the glog function so that eventually we *can* compute its values.

With that in mind, a good starting point in understanding the behavior of glog is to look at a graph. Actually, two different graphical representations provide insight about glog.

First, since glog is defined as the inverse of $f(x) = e^x/x$, a graph of glog is easily obtained by reflecting the graph of f in the line $y = x$, as shown in Fig. 9.1. For later reference, the ln function is also included in the graph.

The graph reveals at once some of the features of glog. For one thing, glog is not a function because for $x > e$ it has two positive values. For $x < 0$ glog is well defined (and negative) but it is not defined for $0 \leq x < e$. When we need to distinguish between glog's two positive values, we will call the larger $glog_+$ and smaller $glog_-$. As suggested by the graph, $glog_-(x) \leq \ln(x) \leq glog_+(x)$ for all $x \geq e$, with equality when $x = e$.

These inequalities appear obvious from the graph, but of course they can be derived analytically. For example, suppose $y = glog\, x$ is greater than 1. By definition, $e^y/y = x$ so $e^y = xy > x$. This shows that $y > \ln x$.

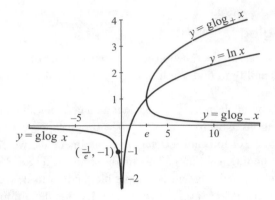

Figure 9.1. Graphs of $y = \text{glog } x$ and $y = \ln x$.

The second graphical representation depends on the fact that glog c is defined as the solution to $e^x = cx$. The solutions to this equation can be visualized as the x-coordinates of the points where $y = e^x$ and $y = cx$ intersect (Fig. 9.2). That is, glog c is determined by the intersections with the exponential curve of a line of slope c through the origin.

This provides additional insight about where glog is defined or single-valued. When $c < 0$ the line has a negative slope and there is a unique intersection with the exponential curve. Lines with small positive slopes do not intersect the exponential curve at all, corresponding to values of c with no glog. There is a least positive slope at which line and curve intersect, that intersection being a point of tangency. It is easy to see that this least slope is $c = e$ and that the point of tangency occurs at $x = 1$. For greater slopes there are two solutions to (2).

In analogy with the natural logarithm, glog obeys a few fundamental identities that follow directly from the definition:

$$\text{glog}(e) = 1,$$

$$\text{glog}(-1/e) = -1,$$

$$\frac{e^{\text{glog } x}}{\text{glog } x} = x,$$

$$\text{glog}\left(\frac{e^x}{x}\right) = x.$$

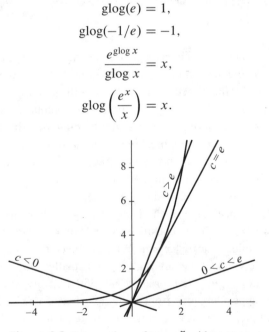

Figure 9.2. Intersections of $y = e^x$ with $y = cx$.

Actually, the last identity is not quite correct. For $x > 1$ the identity should be stated using glog_+ and for $0 < x < 1$ it should be glog_-. But it is easier to remember the ambiguous version, and affix the appropriate subscript when needed.

In addition, there are two more identities that will be useful later. To get the first, rearrange $e^{\text{glog}(x)}/\text{glog}(x) = x$ as

$$e^{\text{glog}(x)} = x\,\text{glog}(x).$$

Take the natural log of both sides to obtain the second:

$$\text{glog}(x) = \ln(x) + \ln(\text{glog}(x)).$$

There are also glog identities roughly analogous to the logarithmic power and product laws, neither very attractive nor likely to be very useful. They are stated below, with verification left as an exercise.

$$r\,\text{glog}\,a = \text{glog}\left(a^r \cdot \frac{(\text{glog}\,a)^{r-1}}{r}\right)$$

$$\text{glog}(a) + \text{glog}(b) = \text{glog}\left(ab \cdot \left[\frac{1}{\text{glog}(a)} + \frac{1}{\text{glog}(b)}\right]^{-1}\right).$$

9.3   Solving Exponential-Linear Equations

By its definition, glog provides immediate solutions to equations of the form $e^x = cx$. Can it also be used to solve other exponential-linear equations? Indeed it can. For example, consider

$$5^{x-3} = 2x + 6. \tag{3}$$

In order to apply glog, this equation must be rewritten in the form $e^u = cu$ for appropriate c and u. First express the right-hand side in the form $2(x + 3)$ and introduce the new variable $u = x + 3$. Then substituting $x = u - 3$ in (3) produces

$$5^{u-6} = 2u.$$

Multiplying both sides by 5^6 leads to

$$5^u = 2(5^6)u,$$

which is almost in the form we want. Since we need the exponential function to appear with base e, replace 5 with $e^{\ln 5}$ to obtain

$$e^{(\ln 5)u} = \frac{2 \cdot 5^6}{\ln 5}(\ln 5)u,$$

or, letting $v = (\ln 5)u$,

$$e^v = \frac{2 \cdot 5^6}{\ln 5}v.$$

From this equation, we observe by inspection that

$$v = \text{glog}\left(\frac{2 \cdot 5^6}{\ln 5}\right).$$

Returning to the original variable, $v = (\ln 5)u = (\ln 5)(x + 3)$, so the equation becomes

$$(\ln 5)(x + 3) = \text{glog}\left(\frac{2 \cdot 5^6}{\ln 5}\right).$$

At last, we can solve for x :

$$x = -3 + \frac{1}{\ln 5}\text{glog}\left(\frac{2 \cdot 5^6}{\ln 5}\right).$$

Notice that $2 \cdot 5^6 / \ln 5$ is a big number, certainly greater than e. Thus, there are two values for $\text{glog}(2 \cdot 5^6 / \ln 5)$, giving two solutions x. Using methods to be presented later, the glog values are found to be, to seven decimal digits,

$$\text{glog}_+ (2 \cdot 5^6 / \ln 5) = 12.39085 \quad \text{and} \quad \text{glog}_- (2 \cdot 5^6 / \ln 5) = 5.150467 \cdot 10^{-5}.$$

These give the solutions of (3) as 4.69887 and -2.99997, approximately.

Similar methods can be applied to any exponential-linear equation

$$AB^{Cx+D} = Px + Q,$$

by first reducing it to the simpler form

$$a^x = b(x + c). \tag{4}$$

The solution to this equation is derived as follows. Begin with a few algebraic rearrangements:

$$a^x = b(x + c)$$
$$a^{x+c} = ba^c (x + c)$$
$$e^{(\ln a)(x+c)} = \frac{ba^c}{\ln a}(\ln a)(x + c).$$

The last equation is in the form $e^y = zy$. Thus

$$(\ln a)(x + c) = \text{glog}\left(\frac{ba^c}{\ln a}\right),$$

from which we obtain

$$x = \frac{1}{\ln a}\text{glog}\left(\frac{ba^c}{\ln a}\right) - c. \tag{5}$$

This makes sense only if $ba^c / \ln(a)$ is less than 0 or greater than or equal to e. In the former case there is a unique solution, while in the latter there are two solutions. Thus glog tells whether (4) is solvable, and gives the solutions when it is.

As a special case, take $c = 0$ in (4). Then we find that

$$a^x = bx \quad \text{iff} \quad x = \frac{1}{\ln a}\text{glog}\left(\frac{b}{\ln a}\right). \tag{6}$$

This is a generalization of the change of base formula for logarithms. Indeed, replacing e with an arbitrary base a, the inverse of the function a^x / x defines what we might reasonably refer to as glog in the base a. This gives the solution(s) of the equation $a^x = bx$. Accordingly, (6) becomes

$$\text{glog}_a (b) = \frac{1}{\ln a}\text{glog}\left(\frac{b}{\ln a}\right), \tag{7}$$

standing as an analog of the corresponding logarithmic identity

$$\log_a(b) = \frac{1}{\ln a}(\ln b).$$

9.4   Applications

The motivation for defining glog was to solve exponential-linear equations, which arise naturally in one type of discrete growth model. As mentioned earlier, glog is closely related to the Lambert W function. In fact, each can be defined in terms of the other. Thus, glog can be used effectively in any application of W, including enumeration of trees, iterated exponentiation, a jet fuel model, and an enzyme kinetics problem, among others. To get the flavor of some applications, we will look at three examples: the previously mentioned discrete growth model, estimation of partial sums of infinite series, and analytic solutions to a variety of equations.

Oil Reserves Model.   Equation (1) originally appeared in a simple discrete model for world petroleum reserves [78, page 277]. Its solution tells when, according to the model, the oil reserves will be exhausted. The model was presented to illustrate difference equation methods, and not as a serious attempt to predict the future availability of petroleum. It shows that accumulating the effects of shifted exponential growth always leads to an exponential-linear equation.

The general formulation involves a discrete model for resource consumption. Time is partitioned into discrete intervals, and c_n represents the amount of the resource consumed in the nth interval. In the example, c_n represents the world consumption of petroleum in year n of the model. The cumulative consumption over n time intervals is then

$$C_n = \sum_{k=1}^{n} c_k.$$

A natural question is: When will the cumulative consumption reach some predefined level L? For the petroleum model, using L as the world reserves at the start of year 0, the question becomes, when will the supply of petroleum be used up?

Now suppose the annual consumption c_n follows a shifted exponential curve. If

$$c_n = ab^n + d,$$

then the cumulative consumption has the form

$$C_n = \alpha n + \beta + \gamma b^n,$$

where α, β, and γ are constants determined by a, b, and d. Specifying a value for C_n thus gives rise to an exponential-linear equation in n. That is how (1) arose.

To solve (1), first put it into the form

$$1.6^x = -122.35(x - 41.31).$$

Then from (5) the solution is

$$x = \frac{1}{\ln 1.6}\text{glog}\left(\frac{-122.35 \cdot 1.6^{-41.31}}{\ln 1.6}\right) + 41.31.$$

Completing the solution requires us to compute $\mathrm{glog}(-9.623 \cdot 10^{-7})$. Because its argument is negative, the glog is uniquely defined and approximately equal to -11.4187. This leads to a final answer of $x = 17.015$. It is to be hoped that this model vastly overestimates consumption, underestimates the reserves, or both!

Estimating Infinite Sums.   A second application concerns estimating the limits of infinite sums. In [20], truncation error estimates $E(n)$ are derived for a number of series. In order to determine n, the number of terms required to assure an error less than ϵ, the inequality $E(n) < \epsilon$ must be inverted. In one instance,

$$E(n) = \frac{1 + \ln(n + \frac{1}{2})}{n + \frac{1}{2}}.$$

This problem is solved by inverting $y = x / \ln x$. That is, we have to solve equations of the form $x = c \ln x$. But with $u = \ln x$, this equation becomes $e^u = cu$. We see immediately that $u = \mathrm{glog}\, c$, showing that $x = e^u = e^{\mathrm{glog}\, c}$.

More Equations.   We have seen that glog can be used to solve any exponential-linear equation. But it can be used to solve a surprising number of other kinds of equations as well. This is hinted at by the preceding application. As another example, glog can be used to give a closed form expression for the iterated exponential function $h(x) = x^{x^{x^{\cdot^{\cdot^{\cdot}}}}}$.

Starting from $h = x^h$, rewrite the right-hand side as $e^{h \ln x}$ to obtain

$$\frac{1}{\ln x} = \frac{e^{h \ln x}}{h \ln x}.$$

Invoking the definition of glog now leads directly to $h \ln x = \mathrm{glog}(1/\ln x)$. Therefore

$$h = \frac{1}{\ln x} \mathrm{glog}\left(\frac{1}{\ln x}\right).$$

In a similar way, glog can be used to solve the equation $x + y = x^y$ for y:

$$y = \mathrm{glog}_x(x^x) - x$$

where the base x glog is as defined earlier. Using (7), we can then derive

$$y = \frac{1}{\ln x} \mathrm{glog}\left(\frac{x^x}{\ln x}\right) - x.$$

Using glog it is possible to solve a variety of other equations that can be transformed into exponential-linear form. The following are examples:

$$a^x = \frac{b}{x + c}$$
$$\ln(bx + c) = px + q.$$

Finding the solutions using glog is left as an exercise.

9.5 Inequalities and Estimates

Earlier, we saw that for $x > e$,

$$0 < \text{glog}_-(x) < \ln(x) < \text{glog}_+(x).$$

This is one of a number of inequalities that relate glog to other functions. For example, using the Taylor expansion, we see that for positive y, $e^y > y^{k+1}/(k+1)!$. This implies that $e^y/y > y^k/(k+1)!$. Now let $y = \text{glog}_+(x)$. Then $e^y/y = x$ and $y^k = [\text{glog}_+(x)]^k$, so the inequality becomes

$$x > \frac{[\text{glog}_+(x)]^k}{(k+1)!}$$

and leads to

$$\sqrt[k]{(k+1)!x} > \text{glog}_+(x).$$

This inequality can be used to show that, like the logarithm, the upper branch of glog grows more slowly than any root function. That is, for any k, $\text{glog}_+(x)/\sqrt[k]{x} \to 0$ as $x \to \infty$. To establish this, compare $\text{glog}_+(x)/\sqrt[k]{x}$ with $\sqrt[2k]{(2k+1)!x}/\sqrt[k]{x}$, holding k fixed and letting x increase without bound.

Although this provides information about the growth rate, it is not of much use in estimating values of the glog function, because it relies on the crude estimate: $e^y > y^k/k!$. A better estimate can be derived using $e^y > ay^k$, with a as large as possible. This requires a point of tangency between $f(y) = ay^k$ and $g(y) = e^y$, as illustrated in Fig. 9.3.

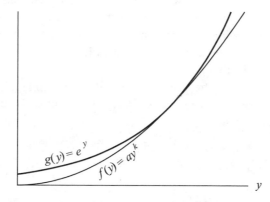

Figure 9.3. Point of tangency between e^y and ay^k.

That is easy to find: require that $f(y) = g(y)$ and $f'(y) = g'(y)$ hold simultaneously. In other words, solve the system

$$ay^k = e^y$$
$$aky^{k-1} = e^y.$$

Clearly, these equations hold only if $y = k$ and $a = e^k/k^k$. That means that $e^y \geq (e/k)^k y^k$ with equality just at $y = k$. Because of the tangency condition, e^y is very close to $(e/k)^k y^k$ for y near k.

We can now derive an estimate for glog. First, divide the inequality by y:

$$e^y/y \geq (e/k)^k y^{k-1}.$$

Take $y = \text{glog}_+(x)$, so $e^y/y = x$ and $y^{k-1} = [\text{glog}_+(x)]^{k-1}$, leading to

$$x \geq (e/k)^k [\text{glog}_+(x)]^{k-1}.$$

Finally, solving for glog,

$$\sqrt[k-1]{\left(\frac{k}{e}\right)^k x} \geq \text{glog}_+(x) \tag{8}$$

with equality at $x = e^k/k$. This inequality again bounds glog_+ using a root function, but it also provides a very good estimate near $x = e^k/k$. In particular, taking $k = 2$ leads to

$$\text{glog}_+(x) \leq \left(\frac{2}{e}\right)^2 x,$$

which implies

$$\text{glog}_+(x) < x.$$

This is not apparent in Fig. 9.1 due to dissimilar scales for the two axes.

9.6 Differentiation and Integration

Generally speaking, travelers to the Calculusian Republic experience an almost overpowering urge to differentiate and integrate any new function that appears on the scene. Up until now, I have resisted temptation, focusing on other aspects of glog lore. But let us resist no longer. Here is a new function, proposed for inclusion in the elementary mathematical canon. What about its derivative and integral?

Differentiating glog follows the standard pattern for differentiating inverse functions. Suppose $y = \text{glog } x$. Then $x = e^y/y$, or $e^y = xy$. Differentiate both sides with respect to x and solve for y', producing

$$y' = \frac{y}{e^y - x}.$$

But we already know that $e^y = xy$, so

$$y' = \frac{y}{xy - x}.$$

Thus, the derivative of glog is

$$\text{glog}'(x) = \frac{\text{glog}(x)}{x\,\text{glog}(x) - x}.$$

Integration is, as usual, another story. To find $\int \text{glog}(x)dx$, how might we proceed? Perhaps integration by parts, taking the integrand as $\text{glog}(x) \cdot 1$, and then differentiating glog and integrating 1. That leaves us with

$$\int \frac{\text{glog}(x)}{\text{glog}(x) - 1} dx,$$

which looks no better than the original problem. Not that it mightn't lead somewhere with sufficient cleverness or luck. However, no obvious route suggests itself. What about starting with the derivative formula? By rearranging it, can we get an idea how to integrate glog? Again, nothing immediately springs to mind.

A substitution? Here we can at least get started. Using the inverse nature of glog, let us define $u = \text{glog } x$, which is equivalent to $x = e^u/u$. We know

$$du = \frac{\text{glog}(x)}{x\text{glog}(x) - x}dx$$

and that can be written

$$(x\text{glog}(x) - x)du = \text{glog}(x)\, dx.$$

On the left, replacing $\text{glog}(x)$ by u and x by e^u/u produces

$$\int \text{glog}(x)\, dx = \int e^u - \frac{e^u}{u}\, du. \tag{9}$$

While this looks promising, it leaves the troublesome e^u/u, which no one has yet figured out how to integrate. In fact, something much stronger can be said. It is not just that no one knows what the integral is, but rather that no integral exists. It has been proven that there is no closed form integral of e^u/u in terms of the familiar elementary functions of calculus.

As a matter of fact, there is a well-developed theory (Liouville's theory of integration in finite terms) that tells exactly which functions have closed form integrals, and which do not. Given the amount of space calculus texts devote to indefinite integration, it is surprising that the existence of such a theory is not better known. Be that as it may, the Liouville theory shows that e^u/u is not integrable in terms of elementary functions, which implies that neither is glog. Otherwise, we could integrate the left side of (9), finding $F(x)$, and then use substitution to conclude

$$e^u - \int \frac{e^u}{u}\, du = F\left(\frac{e^u}{u}\right).$$

This would give an elementary integral of e^u/u, which is a contradiction.

As interesting as this is, it doesn't really close the door on integrating glog. No integral of glog exists in terms of the elementary functions. But glog is not an elementary function, so the Liouville theory says nothing about whether glog might be integrable in terms of *itself* and elementary functions.

This point can be made clearer with a related example. Consider the identity

$$\int \frac{\text{glog}(x)}{x}\, dx = \frac{\text{glog}^2(x)}{2} - \text{glog}(x), \tag{10}$$

which can be verified by differentiation. This shows that $\text{glog}(x)/x$ is integrable in terms of elementary functions and glog. Can it be integrated using only elementary functions? In other words, is there some elementary function $f(x)$ which differs from the right-hand side of (10) by a constant? Suppose that were the case. Then we would have

$$\text{glog}^2(x) - 2\text{glog}(x) = 2f(x) + C,$$

or

$$glog(x) = 1 \pm \sqrt{2f(x) + C + 1}.$$

This expresses glog as an elementary function, which is impossible (see Sidebar 9.1). So $glog(x)/x$ cannot be integrated using elementary functions alone, but it can be integrated using elementary functions plus glog.

As the example illustrates, once glog has been added to our tool box, the idea of integrability (in closed form) should also be extended. What we would like to know is, what functions can be integrated using the elementary functions and glog? I do not know whether glog itself is such a function. But there are a number of other functions, like $glog(x)/x$ which are integrable.

Generalizing (10), for $n \geq 1$,

$$\int \frac{glog^n(x)}{x} \, dx = \frac{glog^{n+1}(x)}{n+1} - \frac{glog^n(x)}{n}.$$

This result, which can also be established by differentiation, is handy for integration by parts. Thus,

$$\int glog^n(x) \, dx = \int x \frac{glog^n(x)}{x} \, dx$$

$$= x \left(\frac{glog^{n+1}(x)}{n+1} - \frac{glog^n(x)}{n} \right) - \int \frac{glog^{n+1}(x)}{n+1} - \frac{glog^n(x)}{n} \, dx.$$

Rearranging leads to the recursion

$$\int glog^{n+1}(x) \, dx = \frac{x}{n} \left(n glog^{n+1}(x) - (n+1) glog^n(x) \right) - \frac{n^2 - 1}{n} \int glog^n(x) \, dx.$$

For $n = 1$, the integral on the right side vanishes, to produce

$$\int glog^2(x) \, dx = x(glog^2(x) - 2glog(x)).$$

It is now easy to compute integrals for successive powers of glog, and the following pattern emerges for $n \geq 2$:

$$\int glog^n(x) \, dx = x \left[glog^n(x) + \sum_{k=1}^{n-1} (-1)^k \frac{n(n-2)!}{(n-k-1)!} glog^{n-k}(x) \right].$$

Although this formula has a certain kind of charm, and is straightforward to verify, a specific example does a much better job of conveying the pattern involved. For $n = 5$, we find

$$\int glog^5 x \, dx = x(glog^5 x - 5glog^4 x + 5 \cdot 3glog^3 x - 5 \cdot 3 \cdot 2glog^2 x + 5 \cdot 3 \cdot 2 \cdot 1glog \, x).$$

These results show that many simple functions involving glog are easily integrated. But it is also interesting to consider whether the introduction of glog allows us to integrate functions that we could not integrate before. Are there any elementary functions that can not be integrated using elementary functions alone, but can be integrated using the elementary functions together with glog?

I do not have an example of such a function, but here is a related, if somewhat foolish, example:

$$\int \frac{1}{x} \, dx = \text{glog}(x) - \ln(\text{glog}(x)).$$

The example is foolish because we already know how to integrate $1/x$, and it merely re-states the identity

$$\ln(x) = \text{glog}(x) - \ln(\text{glog}(x)).$$

Foolish or not, it shows that elementary functions can have integrals expressed in terms of glog. This suggests at least the possibility that including glog in our tool box will allow us to integrate some elementary functions that we could not otherwise integrate.

Before leaving this topic, it is worth mentioning that the W function does have a simple integral in terms of itself and elementary functions: $x(W(x) - 1 + 1/W(x))$ (see [34]). Consequently, W permits the integration of a number of differential equations, accounting for several of the applications of W. It may be that glog, too, can be applied to solve differential equations that arise in a natural way, but that must remain a question for future investigation.

9.7   Computation of glog

It is all very well to contemplate the derivatives and integrals of functions involving glog, but we should not lose sight of our original focus, solving exponential-linear equations. For glog to be useful, we must be able to compute its values. Today glog's close cousin, W, is already included in both Maple and Mathematica, and an internet search readily finds code to add W to other software products, such as Matlab and Mathcad. So if you are interested in serious computing, W is an available stand-in for glog.

For the hobbyest, however, something a little more pedestrian might be appreciated. Someday, if glog catches on, you may be able to find it on a button on your calculator. In the meantime, you can add the equivalent to your calculator in the form of a short program. All that is needed is Newton's method and a procedure for choosing an appropriate initial value. The details will be presented next.

Suppose we wish to compute glog(5). That is equivalent to solving the equation $e^t/t = 5$, or more conveniently, $e^t - 5t = 0$. Solutions to this equation can be found rapidly using Newton's method. (For a review of Newton's method, see page 28.) Setting $f(t) = e^t - 5t$, we compute successive approximations with

$$t_{n+1} = t_n - \frac{f(t_n)}{f'(t_n)} = \frac{t_n - 1}{1 - 5e^{-t_n}}. \tag{11}$$

The computation can easily be implemented on a spreadsheet or calculator. Calculators in the TI-8x series have a particularly convenient option for iteration. Say we decide to start the iteration with $t_0 = 4$. Key 4 into the calculator, and hit the enter key. This is shown in the first line of Fig. 9.4, with a gray rectangle indicating where the enter key was pressed. The calculator's response appears on the second line. Next, enter the computation on the right side of (11), but replacing each instance of t_n with $\boxed{\text{2nd ans}}$ (line 3 of the figure). This

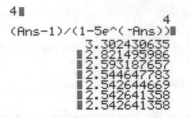

Figure 9.4. Computing glog(5) on a TI calculator.

has the effect of substituting the prior result (4) for t_n. Now hit the enter key repeatedly, shown in the figure by lines that begin with a gray rectangle. Each time the enter key is pressed, the prior calculation is repeated, with the preceding result taking the place of t_n. After six iterations, the process converges — the seventh result is the same as the sixth. Thus we have found glog(5) = 2.542641358. Actually, since this result exceeds e, we have computed $glog_+(5)$.

In this example, we have found glog for a particular value, 5, but the process generalizes easily. To find glog for another value, say 17, all we need to do is change the 5 to a 17 on the right side of (11).

Let us see how Newton's method behaves in general. Mimicking the earlier example, to evaluate glog a, we will solve the equation $f_a(t) = 0$ where

$$f_a(t) = e^t - at.$$

The iteration for computing glog(a) is

$$t_{n+1} = \frac{t_n - 1}{1 - ae^{-t_n}}.$$

Clearly, f_a only has a root if a is in the domain of glog, i. e., if $a < 0$ or $a \geq e$. Depending on which condition holds, the behavior of f_a takes one of the forms illustrated in Fig. 9.5.

For $a < 0$, f_a is increasing everywhere, with a unique root. For $a > e$, f_a has a global minimum at $\ln(a)$ and is monotonic on either side. Since the minimum value is $a - a \ln a < 0$, f_a has two roots separated by $\ln(a)$. For all values of a, f_a is concave up over the entire line.

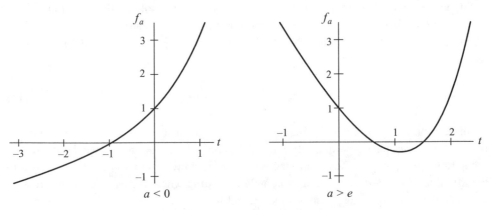

Figure 9.5. Typical graphs for f_a.

These characteristics imply that Newton's method is well behaved for all of the f_a. From any initial value (except $\ln a$ in the case $a \geq e$) Newton's method must converge to a root of f_a, monotonically from at least the second iteration on. If the initial guess t_0 is uphill from the root (that is, $f_a(t_0) > 0$), then the next iterate will be between t_0 and the root, and so all succeeding iterates will move monotonically closer to the root. If t_0 is downhill from the root, then t_1 will be uphill, and the successive iterates will again move monotonically toward the root. When $a < 0$, Newton's method will find the unique root of f_a no matter how t_0 is defined. For $a \geq e$, from any initial value greater than $\ln(a)$ Newton's method will converge to the greater root and from an initial value less than $\ln(a)$ to the lesser root.

Although Newton's method is robust for f_a, computationally it is desirable to select t_0 as close as possible to the root. To do this, it will be helpful to partition the graph of glog into several segments, as indicated in Fig. 9.6. Segments A and C are characterized by large values of $|x|$ and small values of $|\text{glog}(x)|$. On segment B, $|x|$ takes on small values while $|\text{glog}(x)|$ grows without bound. On segment E both x and $\text{glog}(x)$ increase to infinity. Finally, segment D is a neighborhood of the branch point $(e, 1)$ where glog_+ and glog_- coincide. By approximating glog on each of these sections of the graph, we will be able find suitable starting values for Newton's method.

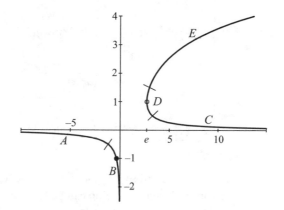

Figure 9.6. Partitioned graph of glog.

Let us begin with segment D. By definition, $y = \text{glog}\, x$ is equivalent to $x = e^y/y$. The second-order Taylor polynomial for e^y at $y = 1$ is $.5e(y^2 + 1)$. Use this in the equation for x :

$$x \approx \frac{.5e(y^2 + 1)}{y},$$

or

$$ey^2 - 2xy + e \approx 0.$$

Solving the quadratic in y leads to

$$\text{glog}\, x = y \approx \frac{1}{e}(x \pm \sqrt{x^2 - e^2})$$

for $x \geq e$. This gives us two values, approximating $\text{glog}_+(x)$ and $\text{glog}_-(x)$. The closer x is to e, the closer y will be to 1, and the more accurate the estimates will be.

A similar approach is effective for segments A and C. There $|y|$ is very small so we can estimate e^y with a quadratic Taylor polynomial expanded about 0. As before $x = e^y/y$ becomes a quadratic equation which can be solved for $y \approx \text{glog } x$.

On segments B and E, we employ a different approach. For segment E, we can use (8), which provides a good estimate for glog x near $x = e^k/k$. Estimating k as the greatest integer in $\ln x$ allows us to determine an estimate for glog anywhere on segment B. This provides a decent estimate for x up to about e^{50}. A similar analysis can be developed for segment B.

Using these estimates to initiate Newton's method, glog can be computed to high accuracy in just a few iterations. By running a large number of numerical experiments, I found approximate optimal transition points between the segments of the graph. Table 9.1 shows how the domain of glog was partitioned and summarizes the convergence results for each segment of the graph in Fig. 9.6. Generally, convergence to about nine or ten decimal digits was observed within three or four iterations, as shown in the table. These results are based on haphazard experimentation; actual performance may vary.

Segment	a Domain	t_0 Formula	Iterations
A	$a < -.42$	$a - 1 + \sqrt{a^2 - 2a - 1}$	3
B	$-.42 \leq a < 0$	$-(k^k e^{-k}/|a|)^{1/k+1}$; $k = \text{int}(\ln|a|) + 1$	2
C	$a \geq 3.4$	$a - 1 - \sqrt{a^2 - 2a - 1}$	2
$D\ (y \geq 1)$	$e < a < 3$	$a/e + \sqrt{(a/e)^2 - 1}$	4
$D\ (y \leq 1)$	$e < a < 3.4$	$a/e - \sqrt{(a/e)^2 - 1}$	3
E	$a \geq 3$	$(k^k e^{-k} a)^{1/k-1}$; $k = \text{int}(\ln a) + 1$	4

Table 9.1. Convergence results and t_0 for glog(a) computation.

The interested reader can explore these results using an Excel spreadsheet available at the website for this book [87]. Alternatively, a glog function can be programmed on a graphing calculator. Details for a program that runs on the TI-83 Plus are provided at the website.

9.8 History, References, and Additional Reading

Most of the material in this chapter was adapted from [81, 82]. The first of these has additional information about methods for computing glog, while the second includes further discussion of integration results.

Although I am unaware of earlier published work about glog, the closely related W function has received considerable attention, with a history that dates back to Euler and Lambert. Where glog is defined by inverting e^x/x, the W function is defined as the inverse of xe^x. Thus, $y = W(x)$ if and only if $x = ye^y$. Writing ye^y as $-1/[e^{-y}/(-y)]$, we obtain

$$W(x) = -\text{glog}(-1/x) \quad \text{and} \quad \text{glog}(x) = -W(-1/x),$$

showing that W can be used in any application of glog, and vice versa.

A complete account of what is known about W can be found in [34], including an impressive list of applications, consideration of numerical and symbolic computation, and an extensive bibliography, as well as a nice historical sketch. Some of this same information is presented in a more abbreviated form in [22]. Good on-line references are [106, 166]. As noted earlier, W is included in both Maple and Mathematica. It is called productlog in the latter.

In considering integration of glog, a brief mention was made of the theory of integration in finite terms. To learn more about this subject, see [93, 115].

The discussion of methods for computing glog on graphing calculators leads naturally to a related question. How do today's calculators actually compute their predefined functions? This is addressed in an interesting paper by Schelin [142].

10

Envelopes and Asymptotes

This chapter is about a pair of topics with a strong visual component. First we will explore the subject of envelopes, which we have seen before in connection with string art patterns and the ladder problem. Then we will look at curves that are asymptotic, a concept with both analytic and geometric interpretations. We will see a surprising condition under which the graphs of functions f and g can be asymptotic, even though $f(x) - g(x)$ is unbounded.

10.1 Envelope Basics

We saw envelopes in string art designs in Chapter 6 and in the solution of the ladder problem in Chapter 7. To review, a family of curves is given by an equation of the form

$$F(x, y, \alpha) = 0 \qquad (1)$$

where, for each value of α the equation defines a curve C_α in x and y. An example from Chapter 6 is shown in Fig. 6.6 on page 126. The equation for that family is

$$\frac{x}{\alpha} + \frac{y}{6 - \alpha} - 1 = 0.$$

Here, α has a geometric interpretation: it is the x-intercept of the line C_α.

As the figure shows, the pattern of lines seems to create an additional curve, in this case marking a boundary of the region swept out by the lines. That boundary curve is called an *envelope* for the family of lines.

The discussion in Chapter 6 also described how to find an equation for the envelope. Differentiate the equation for the family with respect to α, to get a pair of equations

$$F(x, y, \alpha) = 0$$
$$\frac{\partial F}{\partial \alpha}(x, y, \alpha) = 0.$$

Eliminating α produces an equation for the envelope. This procedure is the *envelope algorithm*.

There is some evidence to suggest that the computation of envelopes was once a standard topic in calculus. This is certainly the impression left by [36, 48, 60], all of which date to

193

the 1940s and 1950s. However, anecdotal reports by colleagues who were students and teachers of calculus during that time are inconsistent. In today's calculus texts (or at least, in their indices), one finds no mention of envelopes. The topic is covered in older treatments of calculus [35, 100] and advanced calculus [158] and the expositions in these sources tend to be similar. My informal survey (see page 157) suggests that the topic of envelopes was common in calculus texts throughout the 19th century. Was the topic common enough in the calculus curriculum in the first half of the twentieth century to be considered standard? If so, when and why did it fall out of favor? These are interesting historical questions.

If envelopes have been forgotten in calculus texts, they have not disappeared from the mathematical literature. In some expository publications, examples of which appear at the end of the chapter, one readily finds recent mention of envelopes and the envelope algorithm.

The seasoned traveler in Calculusia will find the topic of envelopes well worth a visit. It is visually appealing, closely connected with the calculus curriculum, and has modern applications as well. As we proceed, I hope you will see why this subject was one of the more popular Caculusian destinations of an earlier era.

10.2   The Traditional Definition of Envelope

Heretofore I have described the envelope of a family of curves as the boundary of the region the curves occupy. Classically, though, an envelope of a family of curves is defined as a curve that is tangent at each of its points to some member of the family. This will be referred to as the *traditional* definition of envelope if we need to distinguish between the two definitions. In many cases both definitions refer to the same curve. This can be seen in earlier figures of envelopes in Chapters 6 and 7, where the boundary curves are in fact tangent at each point to one of the lines in the defining family.

To illustrate the traditional definition, other examples are easily constructed. Given any curve C, consider the family of tangent lines C_α. Clearly, the original curve C is tangent at each of its points to some line in the family, so C is the envelope of the family, according to the traditional definition.

As a particular example, let C have equation $y = x^2$. At $(\alpha, \alpha^2) \in C$, the tangent line C_α has slope 2α and equation

$$y - \alpha^2 = 2\alpha(x - \alpha),$$

or, in the form of (1),

$$2\alpha x - y - \alpha^2 = 0.$$

This is the family of tangent lines to the parabola C, and the parabola, in turn, is the envelope of the family of lines. See Fig. 10.1, where the envelope appears clearly as the boundary of the region swept out by the lines C_α.

If we repeat the construction for $y = x^3$, something different happens. This time we find that tangent line C_α has equation

$$y - \alpha^3 = 3\alpha^2(x - \alpha).$$

For any (x, y) this equation must hold for at least one α. To see this, rewrite the equation as

$$2\alpha^3 - 3x\alpha^2 + y = 0$$

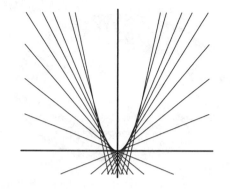

Figure 10.1. A parabola is the envelope of its family of tangent lines.

and recall that every cubic equation has at least one real root. This shows that every point (x, y) is on a tangent line to C for some α, and hence the family of tangent lines sweeps out the entire plane. Therefore, while the original curve C is an envelope of the family of lines in the traditional sense, it is not a boundary for the region swept out by the lines. However, the family of curves for $\alpha \geq 0$ has the right half of C for both an envelope and a boundary, and similarly for the family with $\alpha \leq 0$ and the left half of C.

In both examples, we defined a family of lines made up of all the tangent lines to a given curve, which is then the envelope of the family of lines. We can carry out this construction abstractly. If the original curve C is given by

$$x = f(t)$$
$$y = g(t)$$

then the tangent line at $t = \alpha$ is given by

$$x = f(\alpha) + f'(\alpha)t$$
$$y = g(\alpha) + g'(\alpha)t.$$

Eliminating t leads to

$$xg'(\alpha) - yf'(\alpha) - f(\alpha)g'(\alpha) + g(\alpha)f'(\alpha) = 0,$$

which defines a family of lines, C_α. The original curve is tangent to each C_α at one of its points, and is thus the envelope.

Here is a different example. At each point P of an ellipse E construct a circle C_P that is tangent to the ellipse. Any circle through P whose center is on the normal line to the ellipse at P will do. However the circles are chosen, E is the envelope of the family $\mathcal{F} = \{C_P | P \in E\}$. In Fig. 10.2 $\mathcal{F}$ consists of the circles of unit radius externally tangent to E.

The figure shows that the family of circles sweeps out an oval shaped region. The inside boundary of the region is the original ellipse E. There is another boundary curve on the outside of the region. The two boundary curves in the figure are the only solutions of the envelope algorithm. This leads to the discovery that the outer boundary is *parallel* to the ellipse (in the sense discussed on page 150). Although it has an oval shape, it is not an ellipse, as we shall see presently.

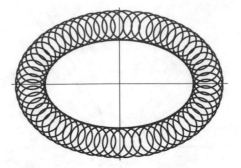

Figure 10.2. An ellipse is an envelope to a family of tangent circles.

Let us use the envelope algorithm for the family of circles $\mathcal{F}$. First we will need to define the family with an equation of the form (1). A member of $\mathcal{F}$ is a unit circle tangent at a point of E, which we assume to be centered at the origin with semi-axes a and b. The tangent point can be expressed parametrically as the vector $\mathbf{r}(\alpha) = (a \cos \alpha, b \sin \alpha)$ where $0 \leq \alpha \leq 2\pi$ (see page 164). The center of the circle is one unit away along the outward normal vector to the ellipse. Since $\mathbf{r}' = (-a \sin \alpha, b \cos \alpha)$ is tangent to the ellipse, an outward normal vector is given by $\mathbf{N} = (b \cos \alpha, a \sin \alpha)$, and the parallel unit vector is $\mathbf{n} = \mathbf{N}/|\mathbf{N}|$. This shows that the center of the unit circle tangent to the ellipse at $\mathbf{r}$ is

$$\mathbf{c}(\alpha) = \mathbf{r}(\alpha) + \mathbf{n}(\alpha). \tag{2}$$

If $\mathbf{p} = (x, y)$ is a point on this circle, then its distance from the center is 1. In terms of vectors, the distance from $\mathbf{p}$ to $\mathbf{c}$ is $|\mathbf{p} - \mathbf{c}| = \sqrt{(\mathbf{p} - \mathbf{c}) \cdot (\mathbf{p} - \mathbf{c})}$. Thus, the equation of the circle can be written

$$((x, y) - \mathbf{c}) \cdot ((x, y) - \mathbf{c}) - 1 = 0. \tag{3}$$

The left side of this equation can be taken as $F(x, y, \alpha)$.

Now we can apply the envelope algorithm. Differentiating with respect to α gives

$$2((x, y) - \mathbf{c}(\alpha)) \cdot (\mathbf{c}'(\alpha)) = 0.$$

Thus, for any point (x, y) of the envelope, $(x, y) - \mathbf{c}$ is perpendicular to $\mathbf{c}'(\alpha)$. Differentiating (2) we get

$$\mathbf{c}'(\alpha) = \mathbf{r}'(\alpha) + \mathbf{n}'(\alpha).$$

Next we observe that both $\mathbf{r}'$ and $\mathbf{n}'$ are perpendicular to $\mathbf{n}$. In the first place, $\mathbf{r}' \perp \mathbf{n}$ by definition. In the second, since $\mathbf{n}$ is a unit vector, $\mathbf{n} \cdot \mathbf{n}$ is constant and its derivative, $2\mathbf{n} \cdot \mathbf{n}'$ vanishes. Thus, since $\mathbf{r}'$ and $\mathbf{n}'$ are both perpendicular to $\mathbf{n}$, they are parallel. This shows that (x, y) is on the envelope if and only if $(x, y) - \mathbf{c}$ is perpendicular to a multiple of $\mathbf{r}'$, and hence parallel to $\mathbf{n}$. Thus, we can write $(x, y) - \mathbf{c} = t\mathbf{n}$ for some real t.

To complete the envelope algorithm, substitute this in (3) to get

$$(t\mathbf{n}) \cdot (t\mathbf{n}) - 1 = 0.$$

It follows that $t = \pm 1$, so the points satisfying the envelope algorithm are $(x, y) = \mathbf{c} \pm \mathbf{n} = \mathbf{r} + \mathbf{n} \pm \mathbf{n}$, that is, either $\mathbf{r}$ or $\mathbf{r} + 2\mathbf{n}$. The points generated by the first possibility comprise the original ellipse E, while those from the second satisfy the definition of the parallel curve to E, two units away in the direction of the outward normal $\mathbf{n}$. Call this curve G.

The curve G has parameterization $\mathbf{r}(\alpha) + 2\mathbf{n}(\alpha)$. This permits us to compute a tangent vector for each of its points. If this were an ellipse, it would have semi-axes $a+2$ and $b+2$, and so satisfy

$$\frac{x^2}{(a+2)^2} + \frac{y^2}{(b+2)^2} = 1.$$

By implicit differentiation, we find the slope of the tangent line at (x, y) to be

$$\frac{dy}{dx} = \frac{-(b+2)^2 x}{(a+2)^2 y}. \tag{4}$$

A little algebra now shows that, unless the original ellipse is a circle, the points $(x(\alpha), y(\alpha))$ given by the parameterization of G do not satisfy (4), and thus G is not an ellipse.

In this example we have used the envelope algorithm to analyze the envelope curves for a family of circles tangent to an ellipse. The algorithm confirmed that one of the envelope curves was the original ellipse, and showed that the other envelope curve was a parallel of the ellipse. Knowing this allowed us to deduce that the second envelope curve was not itself an ellipse. As we carried out the envelope algorithm, none of the vector analysis depended on E being an ellipse. In general, for any smooth curve C, the family of tangent circles of constant radius R has two envelope curves, one being C itself, and the other being a parallel of C at a distance of $2R$.

Envelopes were originally introduced in this book as boundary curves, and the envelope algorithm was justified based on this understanding in Chapter 6. The traditional definition of an envelope curve requires it to be tangent at each point to one of the curves in the family. Must an envelope curve defined in this way still satisfy the conditions of the envelope algorithm? Under suitable assumptions the answer is yes, as we will see next.

10.3   The Envelope Algorithm Rederived

As always, we consider a family of curves $\{C_\alpha\}$ where $C_\alpha = \{(x, y) | F(x, y, \alpha) = 0\}$ for each α. The envelope is a curve C^* tangent at each of its points to one of the curves C_α. We want to show that for each $(x, y) \in C^*$, there exists an α such that F and $\partial F/\partial \alpha$ both vanish at (x, y, α).

In any derivation of this result, some assumptions must be made, e. g. that $F(x, y, \alpha)$ is differentiable with respect to x, y, and α. In classical derivations, one generally also assumes that locally the envelope C^* is smoothly parameterized by α. By definition, each point P of C^* is a point of tangency to some member C_α of the family of curves, and each C_α is tangent to C^* at some point P. This suggests that P can be defined as a function of α. However, some caution is necessary. If a curve C_α touches the envelope in multiple points, there will be ambiguity in defining $P(\alpha)$. This is the situation when we generate the astroid as the envelope of a family of ellipses, as in Fig. 7.6. In this case there are many functions $P(\alpha)$ that map the parameter domain to the envelope, not all of which are continuous. Although in this example it is possible to choose $P(\alpha)$ so as to obtain a smooth parameterization of the envelope, it is not clear how this can be done in general.

Accordingly, in what follows we assume that C^* has a smooth parameterization $P(\alpha) = (x(\alpha), y(\alpha))$ with $P(\alpha)$ a point of intersection of C^* and C_α.

We can then derive the envelope algorithm from the traditional definition of envelope as follows. Any point (x, y) of the envelope is $P(\alpha)$ for some α. Then $(x, y) = (x(\alpha), y(\alpha))$, and F vanishes at (x, y, α). This establishes the first of our desired conclusions.

For the second, since $P(\alpha) \in C_\alpha$, $F(x(\alpha), y(\alpha), \alpha)$ vanishes identically as a function of α. Thus its derivative is zero. Viewing F as a function of three variables, the chain rule gives

$$\frac{\partial F}{\partial x}\frac{dx}{d\alpha} + \frac{\partial F}{\partial y}\frac{dy}{d\alpha} + \frac{\partial F}{\partial \alpha} = 0. \tag{5}$$

We can also view F as a function of two variables, thinking of α as a fixed parameter. Then the xy-gradient of F is normal to the curve C_α, which is a level curve of F. Meanwhile, the parameterization P provides a tangent vector $(\frac{dx}{d\alpha}, \frac{dy}{d\alpha})$ at each point of C^*. At the point $(x(\alpha), y(\alpha))$, C_α and C^* are tangent, so the normal vector $\nabla_{xy} F = (\frac{\partial F}{\partial x}, \frac{\partial F}{\partial y})$ is orthogonal to the tangent vector $(\frac{dx}{d\alpha}, \frac{dy}{d\alpha})$. This shows that the first two terms on the left side of (5) add to 0, and hence $\frac{\partial F}{\partial \alpha} = 0$.

This justifies the envelope algorithm in the context of the traditional definition of envelope. It shows that at each point (x, y) of the envelope there is a value of α for which both $F(x, y, \alpha) = 0$ and $\frac{\partial F}{\partial \alpha}(x, y, \alpha) = 0$. This is a necessary condition, and it can be satisfied by points that are not on the envelope. Indeed, we can construct an example of this phenomenon by reparameterizing the family of curves. The idea behind the construction will be clear from the case of the string art pattern at the start of the chapter, where the family of lines is given by

$$\frac{x}{\alpha} + \frac{y}{6 - \alpha} - 1 = 0. \tag{6}$$

Let $s(\alpha)$ be any differentiable function taking $\mathbb{R}$ onto $\mathbb{R}$. Then the equation

$$\frac{x}{s(\alpha)} + \frac{y}{6 - s(\alpha)} - 1 = 0$$

parameterizes the same family of lines as (6), and so has the same envelope. Applying the envelope algorithm, we compute the partial derivative with respect to α, obtaining

$$\frac{-xs'(\alpha)}{s(\alpha)^2} + \frac{ys'(\alpha)}{(6 - s(\alpha))^2} = 0.$$

This will be satisfied for any value of α where $s'(\alpha) = 0$. If α^* is such a value, then every point of the line

$$\frac{x}{s(\alpha^*)} + \frac{y}{6 - s(\alpha^*)} = 1$$

satisfies the two conditions of the envelope algorithm. That is, the entire line segment corresponding to α^* will be produced by the envelope algorithm. No such line is actually included in the boundary of the string pattern, nor can any such line be tangent to all the lines in the family. This illustrates how the envelope algorithm can produce extraneous results.

It also shows a difficulty with one of our assumptions. Until we have found the envelope for a particular family, how can we tell whether or not it can be smoothly parameterized in terms of α? Moreover, the existence of a smooth parameterization of the envelope may depend on how the family itself is parameterized. For the example, if $s(\alpha) = \alpha$, so that the original parameterization of F is retained, then the envelope curve does have a smooth parameterization $P(\alpha)$. But with $s(\alpha) = \alpha^3$, the parameterization of the envelope becomes $P(\alpha^3)$. This is not smooth at $\alpha = 0$ because its derivative vanishes there.

A more complete discussion of such subtleties can be found in [35] and [138], where conditions giving rise to extraneous results from the envelope algorithm are characterized. As Courant remarks, once the envelope algorithm produces a curve, "it is still necessary to make a futher investigation in each case, in order to discover whether it is really an envelope, or to what extent it fails to be one." Graphing software can often give a clear picture of the envelope of a family of curves, and so guide our understanding of the results of the envelope algorithm.

10.4 Boundary Points On Envelopes

In one of our examples, we saw that an envelope curve, in the traditional sense, does not have to lie on the boundary of the region swept out by the family of curves. But usually if there is a boundary to the region, it must fall along an envelope. To see this, let us consider a family of curves $\{C_\alpha\}$ defined by $F(x, y, \alpha) = 0$, where α is restricted to a closed interval $[\alpha_0, \alpha_1]$. Let Ω be the region that is swept out by the curves. That is, Ω is the union of the curves C_α for $\alpha_0 \leq \alpha \leq \alpha_1$. And let G be a curve on the boundary of Ω.

Since each of the curves C_α is contained within Ω, none of them can cross G. But each point of G must lie on one of the curves. To see this, consider a point $P \in G$, and a sequence of points P_j in Ω converging to P. Each P_j is on C_{α_j} for some $\alpha_j \in [\alpha_0, \alpha_1]$. So there is a convergent subsequence α_{j_k} with limit α^*. By the continuity of F, $P = \lim P_{j_k}$ is a point on C_{α^*}. Since this curve cannot cross G at P, it must be tangent there.

This shows that at each of its points P, G is tangent to one of the curves C_α. That is not quite enough to make G part of the envelope, because it is possible that every point of G comes from the same α. For example,

$$F(x, y, \alpha) = x^2 + y^2 - \sin^2 \alpha, ;\ \ 0 \leq \alpha \leq \pi$$

describes a family of circles centered at the origin with radii varying smoothly between 0 and 1. The region it sweeps out is the closed unit disk $x^2 + y^2 \leq 1$ and the boundary curve is the unit circle. That is not an envelope for the family except in a trivial sense. At each of its points the unit circle is tangent to a member of the family, but it is always the same member, since in this case the boundary curve coincides with one of the members of the family. But when the boundary curve is smoothly parameterized by α, this situation cannot arise. Then the boundary of Ω will fall along an envelope in the traditional sense.

The family of circles has another interesting property. Although the boundary is not an envelope, it is still revealed by the envelope algorithm because the entire curve for $\alpha = \pi/2$ satisfies $\frac{\partial F}{\partial \alpha} = 0$. Here, the fact that the envelope algorithm produces the boundary of the region is consistent with the argument presented in Chapter 6.

As we have now seen, under certain assumptions the boundary of the region swept out by a family of curves will lie on an envelope of that family. In the proof of this result, α was restricted to a closed interval. This assumption allowed us to infer that each point of

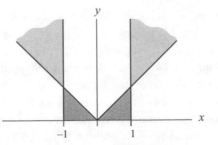

Figure 10.3. The region swept out by curves of the form $y = |x|^\alpha$ for $\alpha \geq 1$. The boundary of the region is not an envelope. Points of the boundary do not, in general, lie on one of the curves.

the boundary lies on some curve of the family. When α is unbounded, this need not be the case. For example, the family with equations $y = (x^2)^\alpha = |x|^{2\alpha}$ for $\alpha \geq 1/2$ sweep out the shaded region in Fig. 10.3, whose boundaries lie on the lines $y = 0$, $y = \pm x$, and $x = \pm 1$, none of which is an envelope for the family of curves.

Another View of the Boundary. Relating the envelope to the boundary also gives an insight about why the envelope technique works. View $F(x, y, \alpha) = 0$ as defining a level surface S of the function F in $xy\alpha$-space. The family of curves can then be viewed as the set of level curves for this surface. The region Ω swept out by the family of curves is the projection of S on the xy-plane. Now suppose A is a point on S that projects to a point P on the boundary of Ω. The tangent plane to S at A must be vertical and project to a line in the xy-plane. Otherwise, there is an open neighborhood of A on the tangent plane that projects to an open neighborhood of P in the xy-plane, and that puts P in the interior of Ω. So the tangent plane is vertical. That implies a horizontal normal vector to S at A, whence the gradient of F, which is normal to S at each point, must be horizontal at A. This shows that the partial derivative of F with respect to α vanishes at A, which is the derivative condition of the envelope algorithm.

10.5   Intersections of Neighboring Curves

As a final topic related to envelopes, I cannot resist mentioning one more way to think about them. The idea is to consider a point on the envelope as the intersection point of two neighboring members of the family. Each α gives us one member of the family of curves, and the intersection of curves for two successive values of α gives a point on the envelope. This is the *limiting position* interpretation of envelope.

Though the idea of successive α's is not literally meaningful, it can be formulated using limits: express the intersection of the curves C_α and $C_{\alpha+h}$ as a function of h and α, and take the limit as h goes to 0. It is instructive to do this for the example in which the curves C_α are tangent lines to $y = x^2$. For this family of lines, $F(x, y, \alpha) = 2\alpha x - y - \alpha^2$. Find the intersection of the lines $F(x, y, \alpha) = 0$ and $F(x, y, \alpha + h) = 0$ and then take the limit as h goes to 0. In the process, not only do you find the equation of the envelope $y = x^2$, but you can also observe echoes of differentiation with respect to α.

We can relate the limiting position idea to the envelope algorithm using an argument like the one for boundary points. Fix α and suppose (x^*, y^*) is the limit of points of intersection of C_α and $C_{\alpha+h}$ as $h \to 0$. We want to show that at (x^*, y^*, α) both F and $\partial F/\partial \alpha$ vanish.

For a decreasing sequence $h_k \to 0$, let (x_k, y_k) be the point of intersection of C_α and $C_{\alpha+h_k}$. Then by assumption $(x_k, y_k) \to (x^*, y^*)$ as $k \to \infty$. Since (x_k, y_k) is on C_α,

$$F(x_k, y_k, \alpha) = 0, \tag{7}$$

so continuity implies that $F(x^*, y^*, \alpha) = 0$. We also know that (x_k, y_k) is on $C_{\alpha+h_k}$, hence $F(x_k, y_k, \alpha + h_k) = 0$. According to the mean value theorem

$$F(x_k, y_k, \alpha + h_k) = F(x_k, y_k, \alpha) + \frac{\partial F}{\partial \alpha}(x_k, y_k, c_k)h_k$$

for some c_k between α and $\alpha + h_k$. Combined with (7), this shows

$$\frac{\partial F}{\partial \alpha}(x_k, y_k, c_k) = 0.$$

Now let $k \to \infty$. Then $(x_k, y_k) \to (x^*, y^*)$ and $c_k \to \alpha$, so by continuity of $\partial F/\partial \alpha$,

$$\frac{\partial F}{\partial \alpha}(x^*, y^*, \alpha) = 0.$$

Thus we have shown that both F and $\partial F/\partial \alpha$ vanish at (x^*, y^*, α), as desired.

We have seen that points determined by the limiting position approach satisfy the conditions of the envelope algorithm. Courant [35] goes the other way, giving a heuristic derivation of the algorithm based on the neighboring curves interpretation. For Rutter [138], the limiting position formulation is one of three closely related but distinct definitions of the envelope concept. He shows by example that a family of curves with an envelope in the traditional sense need have one in the limiting position sense.

To construct Rutter's example, begin with the ellipse

$$\frac{x^2}{a^2} + \frac{y^2}{b^2} = 1$$

and at each point compute the osculating circle. This is the circle whose curvature matches that of the ellipse, and whose center and radius are the center and radius of curvature of the ellipse. The situation is illustrated for an ellipse with $a = 8$ and $b = 4$ in Fig. 10.4, which

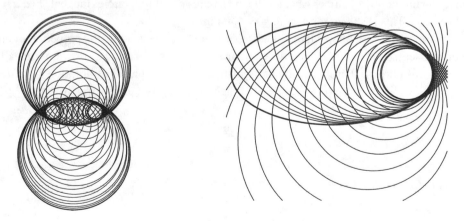

Figure 10.4. Circles of curvature for an ellipse.

shows two different views. On the left the ellipse is shown with tangent circles all around its circumference. The region Ω swept out by the circles is the union of two overlapping disks. On the right is an enlargement showing greater detail for the tangent circles nearest the vertex $(8, 0)$.

By definition, each circle in the family is tangent to the ellipse somewhere, so the ellipse is an envelope. But the ellipse cannot be obtained as the limiting points of intersections of neighboring circles. In fact, the neighboring circles are disjoint! This surprising state of affairs is shown in Fig. 10.5. In the leftmost part of the figure the ellipse and two neighboring circles appear. The center part shows a closeup near the points where the circles touch the ellipse. On the right, the closeup is shown without the ellipse, revealing that the circles are disjoint.

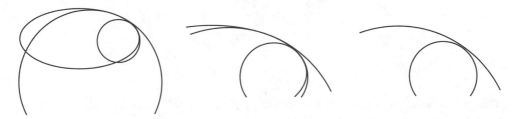

Figure 10.5. Neighboring circles are disjoint.

The same example also exhibits some of the other exceptional behaviors that have been discussed. We can show that the original ellipse does satisfy the two conditions of the envelope algorithm, and so the algorithm would properly identify the envelope for this family of circles. The circles of curvature for each of the vertices $((\pm a, 0), (0, \pm b))$ also satisfy the conditions of the envelope algorithm, although they are *not* part of the envelope. They do contain Ω's boundary, however. This is evident on the left side of Fig. 10.4 where the outer boundary consists of arcs of the circles tangent to E at $(0, b)$ and $(0, -b)$. On the right in Fig. 10.4 the small circle tangent to E at $(a, 0)$ appears as the boundary for a subset of the family of circles. This explains why this small circle and its reflection in the minor axis of the ellipse are visually apparent on the left side of the figure. For other choices of a and b, the full family of circles can have boundary arcs on all four circles tangent at $(\pm a, 0)$ and $(0, \pm b)$. This is shown for $a = 8$, $b = 6$ in Fig. 10.6.

Experimenting with figures like this can contribute to understanding the properties of envelopes discussed above, and I highly recommend it. Modern graphical software is ideally suited to this purpose. For the family of osculating circles, graphical exploration is abetted by the following formulae ([138, p. 192]). If the ellipse is given parametrically by

$$x = a\cos(\alpha)$$
$$y = b\sin(\alpha),$$

then at $(x(\alpha), y(\alpha))$, the radius of curvature is

$$\frac{(a^2\sin^2\alpha + b^2\cos^2\alpha)^{3/2}}{ab}$$

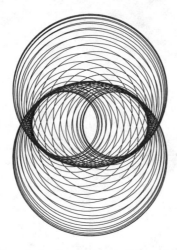

Figure 10.6. Family of osculating circles for an ellipse with semi-axes 8 and 6. The inner boundary of the shaded region lies on arcs of the two smallest circles in the family. The outer boundary lies on arcs of the two largest circles.

and the center of the circle of curvature is

$$\left(\frac{a^2 - b^2}{a} \cos^3 \alpha, \ \frac{b^2 - a^2}{b} \sin^3 \alpha \right).$$

The subject of envelopes is rich, with many interesting facets to explore, analytically and visually. We have seen that the notions of a boundary curve, the traditional envelope, and the limiting position curve are all related to the envelope algorithm. While we have considered only curves defined explicitly by equations in x and y, there is an analogous development for parametric curves that is also interesting. For those wishing to explore envelopes in a visual way, a computer activity at the website for this book [87] makes it easy to view families of lines and their envelopes.

10.6 Asymptotes

From envelopes we turn to another visual subject, asymptotes. This is a well-known part of Calculusia, and most tourists become acquainted with the same basic facts. If $f(x) \to \pm\infty$ as $x \to a$ from either above or below, then the graph of f has a vertical asymptote at $x = a$. Similarly, if $f(x) \to a$ as $x \to \pm\infty$ then the graph has a horizontal asymptote at $y = a$. Asymptotes do not have to be horizontal or vertical, though. A hyperbola with equation $x^2/a^2 - y^2/b^2 = 1$ is asymptotic to the lines $y = \pm(b/a)x$. In general, a line $y = L(x)$ is an asymptote for function f if $f(x) - L(x) \to 0$ as $x \to \pm\infty$.

Here our route departs from familiar turf and follows the lead of a long-ago calculus student I once knew. Tasked to analyze the graph of $f(x) = x^3 - x$, he dutifully found the intercepts, high, low, and inflection points, intervals on which the graph is increasing and decreasing, concave up and concave down. But then he went a step further, adding a dashed line to indicate the graph of $y = x^3$ identified as an asymptote, as shown in Fig. 10.7.

On first examination, this may prompt a skeptical reaction. Is the graph of f asymptotic to the graph of $g(x) = x^3$? Viewed on a graphing calculator the two curves do appear to

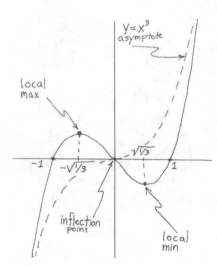

Figure 10.7. A student's analysis of the graph of $f(x) = x^3 - x$.

approach one another, but this is not conclusive. Looking at the question analytically, we see

$$\lim_{x\to\infty} g(x) - f(x) = \lim_{x\to\infty} x = \infty.$$

This suggests that the graphs are not asymptotic. But then what about the graphs of $y = 1/x$ and $y = 1/x^2$? They are both asymptotic to the y-axis and so presumably to each other. However, the limit criterion would make us jump to the opposite conclusion because

$$\lim_{x\to 0+} \frac{1}{x^2} - \frac{1}{x} = \lim_{x\to 0+} \frac{1-x}{x^2} = \infty.$$

We need a better definition of asymptote. Intuitively, asymptotic curves should approach one another as they go to infinity. To formalize this, we could parameterize two curves in the form $\mathbf{r}_1(t) = (x_1(t), y_1(t))$ and $\mathbf{r}_2(t) = (x_2(t), y_2(t))$, and then insist that the distance between $\mathbf{r}_1$ and $\mathbf{r}_2$ go to zero. Since we want this to happen as the curves go to infinity, we will also require that $|\mathbf{r}_1|$ go to infinity. If, say, $|\mathbf{r}_1| \to \infty$ and $|\mathbf{r}_1 - \mathbf{r}_2| \to 0$ as $t \to a$, then the graphs of these curves will be asymptotic.

Unfortunately, this definition is difficult to apply, because it requires the parameterizations to be synchronized. In fact, two different parameterizations of the same curve would not necessarily be asymptotic by this criterion. Consider $\mathbf{r}_1(t) = (t, t)$ and $\mathbf{r}_2(t) = (e^t, e^t)$. As t goes to infinity, both of these parameterizations go to infinity along the line $y = x$, but one goes a lot faster. Thus, $|\mathbf{r}_1 - \mathbf{r}_2|$ does not go to zero.

To avoid this, we might parameterize one of the curves, insisting that as it is traversed, it gets closer and closer to the other curve. Here, we could define the distance from a point $\mathbf{r}(t)$ to a curve C as the distance between $\mathbf{r}(t)$ and the nearest point of C. This too, has a weakness. Suppose that $\mathbf{r}(t) = (t, 0)$ and the curve C has equation $y = \sin(e^t)$. As t goes to infinity, C oscillates more and more rapidly, essentially covering the strip $-1 \le y \le 1$ for large values of x, and so coming close to every point of the graph of $\mathbf{r}$. It is easy to see that the distance from $\mathbf{r}(t)$ to C goes to zero, as follows. Successive x-intercepts of C will be at points $\ln(k\pi)$ and $\ln((k+1)\pi)$. The distance between them is $\ln((k+1)/k)$, which goes to zero as k goes to infinity. For any t, the closest point of C to $(t, 0)$ is no further than the closest x-intercept of C, and the distance from the intercept to $(t, 0)$ is less than the distance between successive x-intercepts. Thus, the distance from $\mathbf{r}(t)$ to C goes to zero as t goes to infinity. But C does not meet our intuitive expectations for an asymptote of the graph of $\mathbf{r}$.

This flaw can be corrected by building symmetry into the definition, requiring that as $t \to \infty$, $\mathbf{r}_1(t)$ gets arbitrarily close to the graph of $\mathbf{r}_2$, and vice versa. But that makes for a rather cumbersome definition. Another alternative is to consider the parts of two curves that lie outside the circle of radius k centered at the origin, and insist that the distance between them goes to zero as k goes to infinity. But then we have to do something about curves that oscillate back and forth between the origin and infinity, like $y = (1/x)\sin(1/x)$. Would we want such a curve to be considered asymptotic to the y-axis?

It seems that satisfactorily defining asymptotic curves is complicated no matter what approach is chosen. That is not to say that no definition is possible, but it shows that there is some subtlety to this subject. We will put aside the quest for a completely general definition, and restrict ourselves to a more concrete approach. We will extend our idea of asymptote to a special case, corresponding to the example of x^3 and $x^3 - x$. We measure the distance between curves like these horizontally, rather than vertically.

To see how this works, consider an arbitrary point $(a, a^3 - a)$ on the graph of $f(x) = x^3 - x$, with $a > 0$. We find the point on the graph of $g(x) = x^3$ that has the same y-coordinate, namely, $a^3 - a$. This point is $(\sqrt[3]{a^3 - a}, a^3 - a)$. See Fig. 10.8. At this value of y, the horizontal distance between the two curves is $h(a) = a - \sqrt[3]{a^3 - a}$. If we can show that this goes to zero as a goes to infinity, it makes sense to consider the two curves asymptotic. Using the horizontal distance in this way is equivalent to asking whether the inverse functions of f and g are asymptotic in the usual sense. The graphs of the inverse functions are the reflections in the line $y = x$ of the graphs of the original functions, and the geometric property of asymptoticity ought to be preserved by reflection. Or, looking at this example another way, interpret the points $(a, a^3 - a)$ and $(\sqrt[3]{a^3 - a}, a^3 - a)$ as parameterizations of the graphs of f and g and apply the earlier ideas about asymptotic parametric curves. Either way, it is certainly reasonable to say that the graphs of f and g are asymptotic, as *curves*, if $h(x) \to 0$ as $x \to \infty$. From now on, we will say that functions f and g are *horizontally asymptotic* when their graphs are asymptotic in this way.

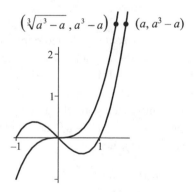

Figure 10.8. Points on the graphs of $f(x) = x^3 - x$ and $g(x) = x^3$ sharing a common y-coordinate.

Let us verify that for $f(x) = x^3 - x$ and $g(x) = x^3$ the horizontal distance $h(x)$ does go to zero. There are several different methods for evaluating $\lim_{x \to \infty} h(x)$. By substituting $1/t$ for x, we could show that the limit is the derivative of a related function, and so use rules of differentiation. Or, using the identity $a - b = (a^3 - b^3)/(a^2 + ab + b^2)$, we could write $h(x)$ in a form that makes it obvious that the limit is zero. The most straightforward approach is probably to use L' Hospital's rule. Write

$$h(x) = \frac{1 - \sqrt[3]{1 - x^{-2}}}{x^{-1}}$$

and both numerator and denominator go to 0 as x goes to infinity. Differentiating the numerator and denominator gives

$$E = \frac{2}{3x \sqrt[3]{(1 - x^{-2})^2}}.$$

Because $E \to 0$ as $x \to \infty$, so does $h(x)$. This confirms that f and g are horizontally asymptotic.

As another example, consider $f(x) = 1/x^2$ and $g(x) = 1/x$. They both have vertical asymptotes on the y-axis, so we would like to be able to say that they are asymptotic to each other. Proceding as before, look at a point $(a, 1/a^2)$ on the graph of f, and the corresponding point $(a^2, 1/a^2)$ on the graph of g. The horizontal distance between the graphs at these points is

$$h(a) = a - a^2.$$

This example is different from the earlier one because here $y \to \infty$ as $a \to 0$ from the right. Therefore, we will conclude that the graphs are asymptotic if

$$\lim_{a \to 0^+} h(a) = 0.$$

Since the limit evidently does equal zero, we happily conclude that the graphs of f and g are asymptotic.

Flushed with these successes, let us now investigate the intuitive idea of that long-ago calculus student. His work suggests a general principle: an asymptote for the graph of a polynomial can be obtained by neglecting all but the highest degree term. Does this idea hold up?

In general, the answer is no. For example, consider $f(x) = (x + 1)^3$ and $g(x) = x^3$. We know that the graph of f is a horizontally shifted copy of the graph of g, so at any point the horizontal distance between these graphs is 1. In this case, neglecting the lower order terms of $f(x) = x^3 + 3x^2 + 3x + 1$ does not produce an asymptote.

Modifying the example, suppose $f(x) = x^3 + x^2$. Is its graph asymptotic to the graph of $g(x) = x^3$? What does your intuition tell you? Have you got a strong hunch?

A typical point on the graph of f is $(a, a^3 + a^2)$ and the corresponding point on the graph of g is then $(\sqrt[3]{a^3 + a^2}, a^3 + a^2)$. Let

$$h(a) = \sqrt[3]{a^3 + a^2} - a.$$

We want to know whether $h(a) \to 0$ as $a \to \infty$.

This time let us use algebraic simplification instead of L'Hospital's rule. Take $b = \sqrt[3]{x^3 + x^2}$ so that $h(x) = b - x$ and $b/x = \sqrt[3]{1 + 1/x} \to 1$ as $x \to \infty$. Using the identity

$$b - x = \frac{b^3 - x^3}{b^2 + bx + x^2}$$

we get

$$h(x) = \frac{x^2}{b^2 + bx + x^2} = \frac{1}{(b/x)^2 + b/x + 1}.$$

Since this has a limit of $1/3$ as $x \to \infty$, the graphs of f and g are not asymptotic.

10.7 Detecting Horizontal Asymptotes

We have now seen two different situations. For $f(x) = x^3 - x$ the graph is asymptotic to $y = x^3$, but for $f(x) = x^3 + x^2$ it is not. Let us develop a way to detect when functions are horizontally asymptotic. Then we will be able to predict whether a polynomial is asymptotic to its leading term.

How can we find the horizontal distance between the graphs of unspecified functions f and g? As in the examples, we begin with a fixed point $(a, f(a))$ on the graph of f. As a simplification, approximate the graph of g near $x = a$ by its tangent line

$$y = g(a) + g'(a)(x - a).$$

Setting $y = f(a)$ in this equation locates a point (x, y) of the tangent line that is on a horizontal line with $(a, f(a))$. This leads to the horizontal distance between $(a, f(a))$ and the tangent line as the absolute value of

$$x - a = \frac{f(a) - g(a)}{g'(a)}.$$

Can we use this as an estimate for the horizontal distance from $(a, f(a))$ to the graph of g?

An answer is suggested by Fig. 10.9, which shows the distance h between the graphs of f and g along a horizontal line. The distance is the difference of x-coordinates x_1 and x_2 for points on the graphs of f and g respectively. They determine two points on each graph, labeled according to the convention $F_k = (x_k, f(x_k))$ and $G_k = (x_k, g(x_k))$.

As shown in the figure both curves are smooth, concave up, and increasing between x_1 and x_2. Under these assumptions, if $f(x_1) > g(x_1)$ then $f(x) > g(x)$ for all x between x_1 and x_2. Indeed, for any such x, we must have $f(x) > f(x_1) = g(x_2) > g(x)$.

The figure also includes the tangent line to the graph of f at F_2 and the tangent line to the graph of g at G_1. The points P and Q are the intersections of the tangent lines with the horizontal line through F_1 and G_2. As the figure shows, $PG_2 < h < F_1Q$. We also can see that the slopes of the tangent lines are given by $f'(x_2) = (f(x_2) - g(x_2))/PG_2$ and $g'(x_1) = (f(x_1) - g(x_1))/F_1Q$. Putting all of these facts together, we reach

$$\frac{f(x_2) - g(x_2)}{f'(x_2)} < h < \frac{f(x_1) - g(x_1)}{g'(x_1)}. \tag{8}$$

Thus, the horizontal distance between the curves can be bounded above and below using tangent line approximations.

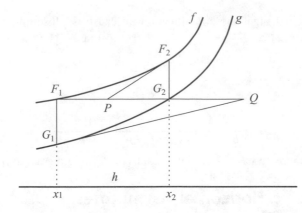

Figure 10.9. Estimating the horizontal distance between the graphs of f and g.

This result can be used see if two functions are horizontally asymptotic, assuming they satisfy the assumptions implicit in Fig. 10.9. Suppose that f and g are concave up and increasing and that $f(x) > g(x)$ in some interval (x_0, ∞). Then, if $(f(x) - g(x))/g'(x) \to 0$ as $x \to \infty$, we can conclude that f and g are horizontally asymptotic. Or, under the same conditions, if $(f(x) - g(x))/f'(x) \not\to 0$ as $x \to \infty$, then f and g are not horizontally asymptotic.

Reasoning in this way, we can derive some general results about conditions for polynomials and rational functions to be horizontally asymptotic, as discussed in the next section. Note though that (8) is not always enough to tell whether a particular pair of functions are horizontal asymptotes. An example in [77] features functions f and g that are horizontally asymptotic but for which (8) is inconclusive, because the lower bound for $h(x)$ goes to zero while the upperbound does not.

10.8   Asymptotic Polynomials and Rational Functions

Let us apply (8) to work out the conditions under which two polynomials are asymptotic. We will consider only the situation for polynomials with positive leading coefficients, because the case of negative coefficients is similar.

Suppose that $f(x) = a_n x^n + a_{n-1} x^{n-1} + \cdots + a_0$ and $g(x) = b_m x^m + b_{m-1} x^{m-1} + \cdots + b_0$ where $a_n > 0$ and $b_m > 0$. For sufficiently large x, the graphs will be concave up and increasing. There are two cases. First, assume that f and g have different degrees. Then for sufficiently large x the polynomial of higher degree will dominate. Therefore, to be consistent with the labeling in Fig. 10.9, assume that $n > m$. Then the degree of $f(x) - g(x)$ is n while that of $f'(x)$ is $n - 1$. Therefore, the first fraction in (8) goes to infinity with x_2, so h goes to infinity. Thus, when f and g have different degrees, they cannot be asymptotic in the horizontal sense.

For the second case, assume that the degrees are the same, so $m = n$. Now the dominant function will be the one with the greater leading coefficient. If $a_n > b_n$ then we can argue as before that the horizontal distance goes to infinity. If $a_n = b_n$, then the degree of $f - g$ is at most $n - 1$, and the degree of both f' and g' will also be $n - 1$. Now we can use (8) to see that the graphs are horizontally asymptotic if and only if $a_{n-1} = b_{n-1}$.

To summarize, we have established the following general result.

Asymptotic Polynomial Theorem: *Two polynomials are horizontally asymptotic if and only if they have the same degree n, equal leading coefficients, and equal coefficients of degree n − 1.*

Although we considered only the case of positive leading coefficients, the theorem is valid for negative leading coefficients as well. One way to see this is to replace f and g by $-f$ and $-g$.

The theorem is consistent with earlier examples. In one we saw that $f(x) = x^3 - x$ is asymptotic to $g(x) = x^3$. In this case, f and g have equal degree $n = 3$, equal leading coefficients of 1, and their terms of degree 2 both have coefficients of 0. Here, the theorem correctly predicts that f and g are horizontally asymptotic.

In a later example we found that $f(x) = x^3 + x^2$ and $g(x) = x^3$ are not asymptotic. Since the second degree coefficients are not equal, this example again agrees with the theorem.

Do you recall the work of the long-ago calculus student? It seemed to suggest a general principle: an asymptote for the graph of a polynomial can be obtained by neglecting all but the highest degree term. Now we can see that this is not quite right. A correct version states that an asymptote for the graph of a polynomial of degree n can be obtained by neglecting all but the terms of degree n and $n - 1$.

The arguments supporting the theorem do not require integral exponents in f and g. We might therefore consider a more general formulation with

$$f(x) = a_1 x^{n_1} + a_2 x^{n_2} + \cdots + a_j x^{n_j}$$

and

$$g(x) = b_1 x^{m_1} + b_2 x^{m_2} + \cdots + b_k x^{m_k}$$

where the exponents are in descending order, the leading coefficients are positive, and n_1 and m_1 both exceed 1, but where we do not assume that the exponents are integers. A modified form of the Asymptotic Polynomial Theorem can then be derived, with the conclusion that f and g are horizontally asymptotic if and only if $n_1 = m_1$ and all coefficients agree for terms of degree $n_1 - 1$ or greater. So $a_1 x^{n_1} + a_2 x^{n_2} + \cdots + a_j x^{n_j}$ is asymptotic to $a_1 x^{n_1}$ if and only if $n_2 < n_1 - 1$. For example, $x^3 + x^{1.999}$ is asymptotic to x^3, but $x^3 + x^2$ is not.

The Asymptotic Polynomial Theorem also has implications for rational functions. It leads to the following result.

Rational Function Principle.   Let $f(x)$ be the rational function

$$f(x) = \frac{x^n + a_{n-1} x^{n-1} + \cdots + a_0}{x^m + b_{m-1} x^{m-1} + \cdots + b_0}$$

where $n > m \geq 1$. Then $f(x)$ is horizontally asymptotic to x^{n-m} if and only if $a_{n-1} = b_{m-1}$.

As justification, use long division to express $f(x)$ as

$$f(x) = q(x) + \frac{r(x)}{x^m + b_{m-1} x^{m-1} + \cdots + b_0},$$

where $q(x)$ is a polynomial of degree $n - m$ and the remainder $r(x)$ is a polynomial of degree less than m. (Here, we assume the remainder is not zero, for otherwise f is a polynomial.) Because $f(x) - q(x) \to 0$ as $x \to \infty$, it suffices to determine the conditions under which x^{n-m} and q are horizontally asymptotic. We know that this occurs if and only if q is missing the term of degree $n - m - 1$. This happens precisely when $a_{n-1} = b_{m-1}$, as carrying out the first step of the long division algorithm will show. This establishes the desired conclusion.

Let us look at a few examples. Suppose we wish to sketch the graph of

$$f(x) = \frac{x^3 + 7x^2 - 3x + 5}{x - 3}.$$

For large x, the highest degree terms will dominate, so it is reasonable to suppose that $f(x)$ is very nearly x^2. However, the graphs of $f(x)$ and x^2 are not asymptotic, as indicated by the Rational Function Principle.

In contrast, for

$$f(x) = \frac{x^3 + 7x^2 - 3x + 5}{x + 7}$$

the Rational Function Principle shows that x^2 is an asymptote (in the horizontal sense).

The difference between the examples is revealed by division. In the first case, we have

$$f(x) = x^2 + 10x + 27 + \frac{86}{x - 3}.$$

Here, the polynomial part is not asymptotic to x^2 because there is a nonzero linear term $10x$. In the second example,

$$f(x) = x^2 - 3 + \frac{26}{x + 7}$$

and the polynomial part is asymptotic to x^2.

10.9   History, References, and Additional Reading

I have not attempted to trace the origins of the envelope concept. It seems to have been a common topic for calculus courses from the 19th into the first half of the 20th centuries. Outside of calculus courses, where else might envelopes be found? The topic appears in works on properties of plane curves (see [138, 171]), but this is a subject that is no longer in the standard mathematics curriculum.

The treatment of envelopes in [171] implies that the topic is properly a part of the study of differential equations. Perhaps it is in this context that envelopes once were considered a standard calculus topic, although that is not the case in [35, 101, 158].

Envelopes continue to appear in expository mathematics articles. See, for example, [16, 56, 112, 143, 147, 148]. There is also an application of envelopes in economics, referred to as the *Envelope Theorem* [45, 165].

The material in this chapter on asymptotes was adapted from my article [77]. That goes into the subject in greater depth, and provides additional results and examples. Regarding asymptotes of rational functions, the results of this chapter shed light on an apparent paradox described in [9, FFF #41]. On related subjects, Bange and Host [7] consider analytic and geometric interpretations of asymptotes in polar coordinates, while Nunemacher [125] discusses linear asymptotes for plane curves of the form $F(x, y) = 0$.

An ambiguity in how asymptotes should be defined has been mentioned in several articles [59, 133, 135]. Here the asymptotes are always lines, and the ambiguity has to do with what it means for a line to be an asymptote of a curve. For example, according to some definitions the curve $y = (\sin x)/x$ is asymptotic to the x-axis, but by other definitions it is not.

11

Derivatives Without Limits

It is truly said that the discovery of calculus, and its subsequent rigorous codification, represents one of the great intellectual achievements of mankind. Primary credit goes to Newton and Leibniz, who independently developed comprehensive and systematic treatments of the subject. However, Newton's and Leibniz's contributions were neither the first word nor the last. Prior to their work, mathematicians knew how to solve many of the main problems of calculus, including finding tangents and areas. And while Newton and Leibniz placed these problems in a coherent context, introducing methods with general applicability, their justifications were not mathematically rigorous. Indeed, significant criticisms were raised almost immediately by their contemporaries. Over the ensuing 150 years, mathematicians continued to worry about the foundations of calculus. These efforts culminated in the work of Cauchy in the 1820s which, according to historian Carl Boyer, "gave to elementary calculus the character that it bears today" [25, p. 514].

Today calculus rests on the limit concept, in terms of which all of the other key concepts of analysis are defined: derivatives, integrals, continuity, infinite series. This approach is so standard that one might easily believe it is the only possibility. But that is not the case. For one example, there is an alternative formulation referred to as nonstandard analysis that legitimates the concept of infinitesimals.

In this chapter we will explore another approach, at least to finding derivatives. The key idea is that two curves are tangent at a point when it comes from a double root of their equation of intersection. This is not a new idea, for it goes back to work of Descartes that predates Newton and Leibniz by about thirty years. Unlike nonstandard analysis, this alternative does not really provide a satisfactory foundation for all of calculus. Nevertheless, it is interesting to see how far Descartes' approach can be extended, and where the limit concept becomes necessary. As we will see, it is possible to develop just about all of the standard material on derivatives found in a first calculus course without using limits.

11.1 Descartes' Method

To introduce Descartes' method, let us begin with an example. Suppose we wish to find the tangent line at $x = 2$ of the polynomial $f(x) = x^4 - 3x^3 + 2x^2 - 3x + 7$. Computing

Is Calculus Correct?

One of the early critics of calculus was George Berkeley (1685 – 1753), Bishop of Cloyne. In 1734 he published a tract entitled *The Analyst* which pointed out flaws in the logical justification for the techniques of calculus. Berkeley did not question the validity of the conclusions reached but he charged that their correctness had not been *proved*. For example, analyzing the calculation of a tangent to a parabola, Berkeley cites two errors. He then states that they cancel out, leaving a correct final answer, but he is not satisfied. After all, as any modern math teacher knows (and has often said), just because you happen to get a correct answer, that doesn't mean your method is valid. In Berkeley's analysis, he attributes the appearance of a correct answer to a fortunate compensation of errors, saying "If you had committed only one error, you would not have come at a true Solution of the Problem. But by virtue of a twofold mistake you arrive, though not at Science, yet at Truth." [13, Section XXII].

The great advantage of calculus over its predecessors was the generality of its methods. In earlier developments, mathematicians had succeeded in finding tangents

Berkeley

$f(2) = 1$, we know the tangent line must pass through the point $(2, 1)$. Therefore, it must satisfy the equation $y - 1 = m(x - 2)$ for some slope m. This shows that the tangent line is the graph of a function

$$L(x) = 1 + m(x - 2).$$

At the point of intersection, we know that $L(x)$ and $f(x)$ agree, independent of the value of m. Consequently, $f(x) - L(x) = 0$ at $x = 2$, and the graph of $f(x) - L(x)$ has an x-intercept there. Since $f(x) - L(x)$ is a polynomial, the root at $x = 2$ implies that it has a factor of $x - 2$. Moreover, experience tells us that the graph is actually tangent to the x axis at $x = 2$ if and only if $(x - 2)^2$ is a factor of $f(x) - L(x)$. We take this as the defining characteristic of the tangent line L. Thus, we wish to find a slope m so that $f(x) - L(x)$ is divisible by $(x - 2)^2$.

Is Calculus Correct? (cont.)

for many specific curves, using methods that varied from one curve to another. In contrast, calculus provided one approach, applicable to all curves. For the known cases, calculus was seen to give correct answers. But that cannot establish the validity of the new methods in general, as Berkeley knew. It is precisely in the application of calculus to new problems, for which the old methods were inadequate, that Berkeley's criticism applies. Without a valid derivation of the method, how can we be sure it gives the right answer?

Nowadays, we have a clear understanding of the methods of calculus and rigorous justifications of their validity. But something curious happened along the way to this understanding: we changed the question. Whereas Berkeley would want to see proof that calculus gives a correct value for the slope of a tangent, the modern formulation recognizes that we have no good way to see if the answer is right or not. Look at it this way. To judge whether a proposed slope is correct, one has to know by independent means what the slope is. If calculus provides the only means for finding the slope, there is no way to prove that the answer is correct.

In the modern development, we use calculus techniques to find a slope, and then *define* the tangent line as the one having that slope. This is reasonable in default of an adequate alternative definition of tangent line. And when an alternative does exist, we can show that the two definitions are in agreement.

Would Berkeley have accepted this viewpoint? Probably not. Berkeley's real issue was the comparison of mathematical truths and articles of religious faith. Today's working mathematician is comfortable with the relativism of mathematical truth. If you define things in such and such a way, you are led to this or that conclusion. Berkeley would have insisted on a more absolute notion of truth. The tangent to a curve is one specific thing, and the goal is to find it. What we call a definition he would have seen as an expression of dogma. In this view, it is simply a matter of faith that calculus gives the *correct* answers.

Algebraically, we express

$$f(x) - L(x) = x^4 - 3x^3 + 2x^2 - 3x + 6 - m(x - 2)$$
$$= (x - 2)\left(\frac{x^4 - 3x^3 + 2x^2 - 3x + 6}{x - 2} - m\right).$$

Carrying out the division on the right (synthetic division works well) we reach

$$f(x) - L(x) = (x - 2)(x^3 - x^2 - 3 - m).$$

As expected, $x - 2$ is a factor of $f(x) - L(x)$.

But we want $(x - 2)^2$ to be a factor. That means $x - 2$ has to be a factor of $T(x) = (x^3 - x^2 - 3 - m)$, or equivalently, that 2 is a root of $T(x)$. Thus, we set $T(2)$ equal to 0. That gives

$$8 - 4 - 3 - m = 0$$

and so $m = 1$.

This is the unique choice of m for which $(x - 2)^2$ is a factor of $f(x) - L(x)$. We conclude that $y = L(x)$ is the tangent line to the graph of f at $x = 2$ if and only if $m = 1$. In particular, $f'(2) = 1$.

Although this seems a little complicated at first exposure, it follows an easily understood pattern. Let us repeat it for a general polynomial $f(x)$, seeking the slope at $x = a$. The tangent line must pass through $(a, f(a))$ and if the slope is m, will have the equation $y - f(a) = m(x - a)$. As before, we interpret the tangent line as the graph of a function

$$L(x) = f(a) + m(x - a).$$

Next, consider

$$f(x) - L(x) = f(x) - f(a) - m(x - a),$$

which we know has a root at $x = a$. Therefore, $x - a$ must be a factor, and we can write

$$f(x) - L(x) = (x - a)\left(\frac{f(x) - f(a)}{x - a} - m\right),$$

knowing that $(f(x) - f(a))/(x - a) = Q(x)$ is a polynomial. Proceeding, we wish $x - a$ to be a factor of $Q(x) - m$, so set $Q(a) - m = 0$. This shows that the slope of the tangent line is given by $f'(a) = Q(a)$.

In summary, we have derived the following algorithm for finding $f'(a)$. Divide $x - a$ into $f(x) - f(a)$ to define a polynomial $Q(x)$, and compute $Q(a)$. This is the same procedure we encountered in the context of Horner's form (see page 11). There, the substitution of a into $Q(x)$ was explained as evaluating the limit of a difference quotient. In the present formulation, we can dispense with the limit concept. By taking divisibility by $(x-a)^2$ as the defining characteristic for a tangent, we derive an algorithm for polynomial differentiation that makes sense with no mention of limits.

Using it we can derive the usual rule for differentiating a polynomial. Begin with an arbitrary polynomial, expressed in the form

$$p(x) = \sum_{k=0}^{n} b_k x^k.$$

Synthetic division of $p(x) - p(a)$ by $x - a$ produces the polynomial

$$Q(x) = \sum_{k=0}^{n-1} c_k x^k$$

where the coefficients c_k follow the pattern we observed in the table on page 10. Thus

$$c_{n-1} = b_n$$
$$c_{n-2} = b_n a + b_{n-1}$$
$$\vdots$$
$$c_0 = b_n a^{n-1} + b_{n-1} a^{n-2} + \cdots + b_1.$$

Since

$$c_k = \sum_{j=k+1}^{n} b_j a^{j-k-1}$$

we get

$$Q(a) = \sum_{k=0}^{n-1} \sum_{j=k+1}^{n} b_j a^{j-k-1} a^k$$

$$= \sum_{k=0}^{n-1} \sum_{j=k+1}^{n} b_j a^{j-1}.$$

Now reversing the order of summation produces

$$Q(a) = \sum_{j=1}^{n} \sum_{k=0}^{j-1} b_j a^{j-1}$$

and since the summand does not depend on k, the inner sum can be evaluated to reach

$$Q(a) = \sum_{j=1}^{n} j b_j a^{j-1}.$$

Thus, since we know $Q(a) = f'(a)$, the familiar rule for the derivative of a polynomial has been established.

The Folium of Descartes.   According to Suzuki [157], Descartes used his method to find tangents to curves of the form

$$x^3 + y^3 = pxy,$$

now known as the folium of Descartes, and set this problem as a challenge to Fermat. This makes an interesting example for us as well.

For concreteness, take $p = 9/2$ and consider the point $(2, 1)$ on the curve, as shown in Fig. 11.1. The tangent line through $(2, 1)$ must have the form

$$y - 1 = m(x - 2)$$

which we express as

$$y = 1 + m(x - 2).$$

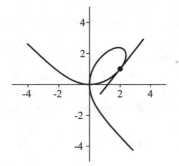

Figure 11.1. Find the tangent line at $(2, 1)$ on the curve $x^3 + y^3 = (9/2)xy$.

This line and the curve

$$x^3 + y^3 = \frac{9}{2}xy$$

intersect at $(2, 1)$. Combining their equations to eliminate y, we obtain

$$x^3 + (1 + m(x - 2))^3 = \frac{9}{2}x(1 + m(x - 2)),$$

which must have a root at $x = 2$. Our goal is to choose m so that $x = 2$ is a double root, and our method will be to divide out a factor of $x - 2$, then set $x = 2$ and solve for m.

We rearrange the equation, expanding the $(1 + m(x - 2))^3$ term and bringing all the nonzero terms to one side. In the process, we preserve all of the $x - 2$ factors, anticipating the factorization step to follow. Thus we reach the equation

$$x^3 + m^3(x - 2)^3 + 3m^2(x - 2)^2 + 3m(x - 2) + 1 - \frac{9}{2}x - \frac{9}{2}xm(x - 2) = 0,$$

and after a further step

$$x^3 - \frac{9}{2}x + 1 + m^3(x - 2)^3 + 3m^2(x - 2)^2 + m(x - 2)(3 - \frac{9}{2}x) = 0. \qquad (1)$$

Next, we are to divide out a factor of $x - 2$. We see that 2 is a root of $x^3 - (9/2)x + 1$ (as it must be), so $x - 2$ divides evenly into that part of the equation, leaving $x^2 + 2x - 1/2$. Dividing (1) by $x - 2$ therefore produces

$$x^2 + 2x - \frac{1}{2} + m^3(x - 2)^2 + 3m^2(x - 2) + m(3 - \frac{9}{2}x) = 0.$$

Setting $x = 2$ reduces this to

$$\frac{15}{2} - 6m = 0.$$

Thus, we find $m = 5/4$. This is the correct slope at $(2, 1)$, as the standard method of implicit differentiation readily verifies.

Repeating this derivation at an arbitrary point (x_0, y_0) of the curve follows essentially the same steps, ultimately producing a general formula for the slope of the tangent line. This is recommended as an exercise for the reader.

Non-polynomial Examples.   Earlier we saw that Descartes' method permits differentiation of all polynomials. As the preceding example hints, the method can be applied more broadly. Rational functions and roots can be treated implicitly, as in the example, but we can also approach these explicitly.

As an example, let us determine the derivative of $f(x) = 8/x$ at $x = 4$. As usual, we begin with a line through the point $(4, f(4)) = (4, 2)$, expressed as the graph of a linear function $L(x) = 2 + m(x - 4)$. We know that $f(x) - L(x) = 8/x - 2 - m(x - 4)$ has a root at $x = 4$, and this is algebraically apparent when we rewrite the equation as

$$f(x) - L(x) = \frac{2(4 - x)}{x} - m(x - 4)$$

$$= (4 - x)\left(\frac{2}{x} + m\right).$$

Recalling the tangency condition, we want $f(x) - L(x)$ to be divisible by $(x - 4)^2$, and that means $m + 2/x$ must be zero when $x = 4$. This is true if and only if $m = -1/2$. Thus, we have shown that $f'(4) = -1/2$.

The general formula for $f'(a)$ can be derived by the same process, using a in place of 4. That leads to $f'(a) = -8/a^2$, which we know is correct.

As a second example, let $f(x) = \sqrt[3]{x}$, and consider a point $(a, \sqrt[3]{a})$ on the graph of f. The tangent line at $x = a$ is the graph of $L(x) = f(a) + m(x - a)$, and as always, $f(x) - L(x) = f(x) - f(a) - m(x - a)$ has a root at $x = a$. Therefore, we factor out $x - a$ to obtain

$$f(x) - L(x) = (x - a)\left(\frac{f(x) - f(a)}{x - a} - m\right)$$

$$= (x - a)\left(\frac{\sqrt[3]{x} - \sqrt[3]{a}}{x - a} - m\right). \tag{2}$$

We want the second factor on the right to have a root at $x = a$, but proceeding as before leads to a new difficulty. Substituting a for x makes the denominator of the fraction zero. Using a familiar algebraic reformulation, though, we obtain

$$\frac{\sqrt[3]{x} - \sqrt[3]{a}}{x - a} = \frac{\sqrt[3]{x} - \sqrt[3]{a}}{\sqrt[3]{x}^3 - \sqrt[3]{a}^3}$$

$$= \frac{\sqrt[3]{x} - \sqrt[3]{a}}{(\sqrt[3]{x} - \sqrt[3]{a})(\sqrt[3]{x^2} + \sqrt[3]{ax} + \sqrt[3]{a^2})}$$

$$= \frac{1}{(\sqrt[3]{x^2} + \sqrt[3]{ax} + \sqrt[3]{a^2})},$$

and that changes (2) into

$$f(x) - L(x) = (x - a)\left(\frac{1}{(\sqrt[3]{x^2} + \sqrt[3]{ax} + \sqrt[3]{a^2})} - m\right).$$

The condition for tangency is that the second factor on the right have a root at $x = a$, so we substitute a for x and set that factor to zero. The resulting equation implies that $m = 1/(3\sqrt[3]{a^2}) = (1/3)a^{-2/3}$.

At this point, the close connection between Descartes' method and the conventional method from calculus should be clear. In Descartes' method we have to substitute $x = a$ into $(f(x) - f(a))/(x - a) - m$, and set the result equal to 0. This requires that we simplify the difference quotient $Q(x) = (f(x) - f(a))/(x - a)$, and then gives the result $m = Q(a)$. In the conventional method, we say we want

$$\lim_{x \to a} \frac{f(x) - f(a)}{x - a}.$$

But in practice, we generally evaluate the limit by first simplifying $Q(x)$ to a form that has no singularity at $x = a$, and then substituting a for x. Procedurally, the two methods are the same, at least for cases where $Q(x)$ can be algebraically simplified. Conceptually, though, they rely on different definitions of tangent lines. Up to this point, the Cartesian formulation works as well as the conventional one while avoiding the limit concept.

11.2 Differentiation Rules

While it is illuminating to compute derivatives by direct application of the definition, that is not a practical approach to differentiation. It is more efficient to establish rules of differentiation for the various ways of combining functions. Then, the derivative of an elementary function is found as a combination of derivatives of simpler functions. In fact, the definition of derivative needs only to be applied to a handful of functions, $x, 1/x, e^x, \sin x$, and the constant function 1. Finding the derivative of anything else is simply a matter of playing by the rules.

With that in mind, it is natural to ask whether the Cartesian approach can be used to derive the sum, difference, product, quotient, and chain rules. The answer is yes.

Before proceeding to derive the rules, we need to formalize a definition of differentiation, based on the preceding examples. Reviewing Descartes' method, we observe a standard algorithm: form the equation $f(x) - L(x) = f(x) - f(a) - m(x - a) = 0$, factor $x - a$ out of the expression in the middle, and select m so that the remaining factor (call it T) has a root at $x = a$. Differentiability thus depends on the existence of the factor T and the constant m.

Can we be sure that $f(x) - f(a) - m(x - a)$ always factors as $(x - a)T(x)$? Formally, we might just define $T(x)$ as $(f(x) - f(a))/(x - a) - m$, and in that sense the factorization is possible. But what sort of function is T? When f is a polynomial, so is $f(x) - f(a) - m(x - a)$, and the factor theorem (page 5) implies that $T(x)$ is also a polynomial. On the other hand, consider $f(x) = |x - a|$, which we know is not differentiable at a. The formal factorization mentioned above gives

$$T(x) = \frac{|x - a|}{x - a} - m$$

which is undefined at $x = a$. Our algorithm fails because we cannot compute $T(a)$, let alone set it to zero. This suggests that we must insist on some restriction on T.

Consider again the example $f(x) = \sqrt[3]{x}$. As we saw earlier, in this example

$$T(x) = \frac{1}{(\sqrt[3]{x^2} + \sqrt[3]{ax} + \sqrt[3]{a^2})} - m.$$

This is not a polynomial, but it is defined and well behaved at $x = a$. Similarly, for $f(x) = 8/x$ we found

$$T(x) = \frac{-2}{x} - m.$$

Here again, though T is not a polynomial, it is defined and well behaved for $x = a$ (except when $a = 0$). These examples suggest that what we want, informally, is for T to be a proper function, not just a formal quotient, and in particular, well behaved at $x = a$.

From our modern perspective, we know that this idea can be formalized using the concept of continuity. The differentiation algorithm will work if there is a continuous function to play the role of T. But how are we to define continuity? In the traditional development of calculus, the definition of continuity depends on the limit concept, which we wish to avoid. In topology, continuity can be defined using open sets, but that seems a long way to go to avoid limits. My preference is to proceed informally, in a manner that would have been acceptable to Descartes, and most likely to critics like Berkeley (see Sidebar 11.1). We will refrain from defining exactly what sort of a function T must be, content to adopt an "I'll know it when I see it" attitude.

With that understanding, the Cartesian approach can be formulated generally. To find a derivative there must be a suitable function T for the factorization $f(x) - f(a) - m(x - a) = (x - a)T(x)$. We can then find the required slope m by insisting that T have a root at $x = a$. These ideas lead to the following definition.

Definition: The function f is said to be differentiable at a if there is a function $T_{f,a}(x)$ and a constant m_f such that the following conditions hold in some neighborhood of $x = a$:

$$(i) \ \ T_{f,a}(x) = \frac{f(x) - f(a)}{x - a} - m_f \ \text{ for } x \neq a$$

$$(ii) \ \ T_{f,a}(a) = 0.$$

In this case, we say that f is differentiable at a, m_f is called the derivative of f at a, and we write $f'(a) = m_f$.

We can now proceed to determine the rules of differentiation. For the simplest functions derivatives can be found directly by applying the definition. But the real power of the definition is in its abstraction, which allows us to formulate rules of differentiation that also apply abstractly. That is, we consider generic functions that satisfy the definition of differentiability and then show that some combination of these functions also satisfies the definition. In this development we echo the general approach of Newton and Leibniz. However, based on Descartes' algorithm, our definition of differentiation avoids (or at least better hides) the uncertain foundations of fluxions and infinitesmals.

The Identity Function and Constants.   To see how the definition works, let us apply it to some particular functions. Consider first $f(x) = x$. The graph of this function is a straight line, so where Descartes' method constructs $f(x) - L(x)$, in this case we will get the constant zero. But does that meet the terms of the definition? To see, let us define $T_{f,a}(x) = 0$ for all x, and $m_f = 1$. Then for $x \neq a$ compute

$$\frac{f(x) - f(a)}{x - a} - m_f = \frac{x - a}{x - a} - 1 = 0 = T_{f,a}(x).$$

This shows that (i) of the definition is satisfied, and (ii) is clearly satisfied as well. So, according to the terms of the definition, we have shown that f is differentiable at a for any a, and that $f'(a) = 1$.

For any constant the graph will again be a straight line, so that $T_{f,a}$ is again the zero function. This time, though, $m_f = 0$. As before, conditions (i) and (ii) defining differentiability are satisfied, justifying the expected result that the derivative of a constant is 0.

The Sum Rule.   Next, let us use the definition to see that the sum of differentiable functions is differentiable and that the derivative of the sum is the sum of the derivatives. We assume that f and g are differentiable at a. That means that there are functions $T_{f,a}$ and $T_{g,a}$ as in the definition. We use them to build an acceptable $T_{h,a}$ for $h(x) = f(x) + g(x)$.

Not surprisingly, we define $T_{h,a}(x) = T_{f,a}(x) + T_{g,a}(x)$ and $m_h = m_f + m_g$. To verify (i), observe that for $x \neq a$,

$$
\begin{aligned}
T_{h,a}(x) &= T_{f,a}(x) + T_{g,a}(x) \\
&= \frac{f(x) - f(a)}{x - a} - m_f + \frac{g(x) - g(a)}{x - a} - m_g \\
&= \frac{(f(x) + g(x)) - (f(a) + g(a))}{x - a} - (m_f + m_g) \\
&= \frac{h(x) - h(a)}{x - a} - m_h.
\end{aligned}
$$

For (ii), we note

$$
T_{h,a}(a) = T_{f,a}(a) + T_{g,a}(a) = 0.
$$

Thus we have shown that h satisfies the definition of differentiability, and $h'(a) = f'(a) + g'(a)$.

The difference rule and the constant multiple rule follow similarly. The details are left as an exercise.

The Product Rule. Once again, we assume that f and g are differentiable at a, and now consider the product $h = fg$. To construct $T_{h,a}$, we begin with the case $x \neq a$. Then

$$
\frac{h(x) - h(a)}{x - a} - m = \frac{f(x)g(x) - f(a)g(a)}{x - a} - m.
$$

We add and subtract $f(a)g(x)$ in the numerator of the fraction (remind you of anything?) to get

$$
\begin{aligned}
\frac{h(x) - h(a)}{x - a} - m &= \frac{f(x)g(x) - f(a)g(x) + f(a)g(x) - f(a)g(a)}{x - a} - m \\
&= g(x)\frac{f(x) - f(a)}{x - a} + f(a)\frac{g(x) - g(a)}{x - a} - m \\
&= g(x)(T_{f,a}(x) + m_f) + f(a)(T_{g,a}(x) + m_g) - m,
\end{aligned}
$$

where the last formulation uses the fact that $T_{f,a}$ and $T_{g,a}$ satisfy condition (i) of the definition. This suggests the proper form for $T_{h,a}$, but we need to choose the value for m so that $T_{h,a}(a) = 0$. Substituting a for x in the last equation above produces $g(a)m_f + f(a)m_g - m$, which vanishes when $m = g(a)m_f + f(a)m_g$. Thus, if we define $m_h = g(a)m_f + f(a)m_g$ and

$$
T_{h,a}(x) = g(x)(T_{f,a}(x) + m_f) + f(a)(T_{g,a}(x) + m_g) - m_h,
$$

both conditions of the definition will be satisfied. Thus, $h(x) = f(x)g(x)$ is differentiable at a and $h'(a) = g(a)f'(a) + f(a)g'(a)$.

The Reciprocal and Quotient Rules. The quotient rule can be established similarly. But we can avoid a direct appeal to the definition by first establishing the chain rule. That will be needed in any case, and is the last rule we'll derive. We'll get to it next. Anticipating the result, here is an efficient way to derive the quotient rule. As in the earlier example with $f(x) = 8/x$, use the definition of differentiation to derive $(1/x)' = -1/x^2$. Next the chain rule gives us the reciprocal rule $(1/g)' = -(g'/g^2)$. Finally, the quotient rule can be derived by expressing f/g as $f \cdot 1/g$ and using the product rule.

The Chain Rule.   For the chain rule, we assume that f is differentiable at a, and that g is differentiable at $b = f(a)$. This implies the existence of the functions $T_{f,a}(x)$ and $T_{g,b}(x)$ specified in the definition of differentiability. For $h(x) = g(f(x))$ we define $m_h = m_f m_g$ and

$$T_{h,a}(x) = (T_{g,b}(f(x)) + m_g)(T_{f,a}(x) + m_f) - m_h$$

and verify that conditions (i) and (ii) hold.

For condition (i) suppose that $x \neq a$. Then either $f(x) = f(a)$ or $f(x) \neq f(a)$. In the former case we have $f(x) = b$ and $T_{f,a}(x) = (f(x) - f(a))/(x - a) - m_f = -m_f$. Therefore

$$T_{h,a}(x) = (T_{g,b}(f(x)) + m_g)(T_{f,a}(x) + m_f) - m_h$$
$$= (T_{g,b}(b) + m_g)(-m_f + m_f) - m_h$$
$$= -m_h.$$

Also,

$$\frac{h(x) - h(a)}{x - a} - m_h = \frac{g(f(x)) - g(f(a))}{x - a} - m_h$$
$$= \frac{g(b) - g(b)}{x - a} - m_h$$
$$= -m_h.$$

This shows that $T_{h,a}(x) = (h(x) - h(a))/(x - a) - m_h$, verifying (i) for the case $f(x) = f(a)$.

We still have to verify (i) when $f(x) \neq f(a)$. In that case we can write

$$\frac{h(x) - h(a)}{x - a} - m_h = \frac{g(f(x)) - g(f(a))}{f(x) - f(a)} \frac{f(x) - f(a)}{x - a} - m_h$$
$$= \frac{g(f(x)) - g(b)}{f(x) - b} \frac{f(x) - f(a)}{x - a} - m_h. \tag{3}$$

Because f is differentiable at a and $x \neq a$,

$$\frac{f(x) - f(a)}{x - a} = T_{f,a}(x) + m_f.$$

Likewise, differentiability of g at b and $f(x) \neq b$ imply

$$\frac{g(f(x)) - g(b)}{f(x) - b} = T_{g,b}(f(x)) + m_g. \tag{4}$$

Substituting these into (3) leads to

$$\frac{h(x) - h(a)}{x - a} - m_h = (T_{g,b}(f(x)) + m_g)(T_{f,a}(x) + m_f) - m_h$$
$$= T_{h,a}(x).$$

Thus, the definitions of $T_{h,a}(x)$ and m_h satisfy condition (i).

For condition (ii), substitute $x = a$ into the definition of $T_{h,a}$ to find

$$T_{h,a}(a) = (T_{g,b}(f(a)) + m_g)(T_{f,a}(a) + m_f) - m_h$$
$$= (T_{g,b}(b) + m_g)(T_{f,a}(a) + m_f) - m_h.$$

But by definition, $T_{g,b}(b) = T_{f,a}(a) = 0$. Therefore

$$T_{h,a}(a) = m_g m_f - m_h = 0.$$

This shows that condition (ii) is satisfied, and hence h is differentiable at $x = a$ with derivative $m_g m_f = g'(f(a))f'(a)$.

Did you notice a flaw in the proof? We assumed that $g(x)$ is differentiable at b, so we can assume that (i) holds in a neighborhood of b. For (4), how can we be sure that $f(x)$ is in that neighborhood? If we could prove that differentiability implies continuity there would be no difficulty. But we agreed to work with an informal definition of continuity in order to avoid the limit concept. In fact, in our definition of differentiability we have only a vague idea that $T_{f,a}$ is well behaved. We could strengthen that, requiring $T_{f,a}$ to be bounded in a neighborhood of a. Then you can prove that $f(x)$ has to be in a neighborhood of $f(a)$, and save the proof of the chain rule. The argument might have satisfied Berkeley and the other critics of Newton and Leibniz, but it brings us perilously close to the limit concept. No matter. Take it that we have pinpointed one place where the limit is indispensible. We will see another before we are done.

Algorithmic Differentiation. The complete set of rules for differentiation requires the definition of differentiability in just a few places. We need to compute directly the derivatives of x, $1/x$, and constant functions. We also need to establish the sum, product, and chain rules. Once those results are in hand, the other rules of differentiation can be obtained as corollaries. We can derive the derivative of a difference $f - g$ by expressing it in the form $f + (-1)g$ and using the sum and product rules. The reciprocal rule is obtained using the derivative of $1/x$ and the chain rule. The quotient rule then follows from the product rule and reciprocal rule.

The chain rule also justifies the standard techniques of implicit differentiation and in particular gives derivatives of inverse functions. Thus, with these few results, we can algorithmically differentiate any algebraic expression, and all without using the limit concept.

To complete the standard development of differentiation found in introductory calculus courses, we need methods to differentiate exponential and trigonometric functions. Here, again, some sort of limit concept seems to be necessary. However, we can still get quite far operating a bit informally. Let us proceed next to examine exponential functions.

11.3 Exponential Functions

We again adopt an informal view of functions and tangents. In particular, we consider it self-evident that an exponential function has a tangent line at each point. Thus for constant $b > 0$, the function b^x is assumed to be differentiable at $x = 0$, with derivative m_b. Initially we do not know the value of m_b. Nevertheless, we can proceed to show that the derivative of b^x is $m_b b^x$.

Let $f(x) = b^x$. At $x = a$, we wish to verify that $f'(a) = m_b b^a$. Because we assume $f'(0) = m_b$, there is a function $T_{f,0}(x)$ satisfying the two conditions of the definition of differentiability. Thus, $T_{f,0}(0) = 0$, and for $x \neq 0$,

$$T_{f,0}(x) = \frac{f(x) - f(0)}{x - 0} - m_b$$

$$= \frac{b^x - 1}{x} - m_b.$$

Let us define $m_f = m_b b^a$ and $T_{f,a}(x) = b^a T_{f,0}(x - a)$. Then for $x \neq a$, because $x - a \neq 0$ we see

$$T_{f,a}(x) = b^a T_{f,0}(x - a)$$

$$= b^a \frac{b^{x-a} - 1}{x - a} - b^a m_b$$

$$= \frac{b^x - b^a}{x - a} - b^a m_b.$$

This shows that condition (i) holds. On the other hand, if $x = a$ then $x - a = 0$, and in this case

$$T_{f,a}(a) = b^a T_{f,0}(0) = 0,$$

satisfying (ii). Thus we have shown that $f(x)$ is differentiable at $x = a$, and that $f'(a) = m_f = m_b b^a$.

This shows that the existence of a tangent line at $x = 0$ implies differentiability for all x. To use this numerically, we need a way to estimate the slope of b^x at $x = 0$. For this purpose, we might use slopes of secant lines. Informally, it appears that increasingly accurate estimates should result from secant lines over shorter and shorter intervals. So long as we only claim to be approximating the slope of the tangent line, we can still avoid taking a limit.

But now let us take a different tack. Geometrically, it is apparent that m_b increases with b. Estimating slopes m_b for various values of b we can tell that some m_b are less than 1 and some are greater than 1. Thus, there must be a particular value of b for which $m_b = 1$. We define that value to be e. Then, because $m_e = 1$, we know that $(e^x)' = e^x$. Note, however, that this definition does not provide much help in estimating e numerically.

From this point forward, we find ourselves on the Calculusian equivalent of an interstate highway. The route is swift and well traveled. Define $\ln x$ as the inverse function of e^x. Then the identity $b = e^{\ln b}$ shows that $b^x = e^{x \ln b}$. Differentiating using the chain rule gives

$$(b^x)' = \ln b \cdot b^x.$$

Now we have a different formulation for the constant m_b described earlier as the slope of b^x at $x = 0$. But we still need a way to compute it.

By implicit differentiation, $(\ln x)' = 1/x$. Assuming the fundamental theorem of calculus, we can write

$$\ln b = \int_1^b \frac{1}{x}\, dx,$$

(since we know $\ln 1 = 0$). With upper and lower estimates for the integral, we can estimate $\ln b$ to whatever accuracy we wish. However, to justify this approach, we need a definition of integration and a proof of the fundamental theorem. These are not inconsiderable requirements, and are beyond the scope of the present discussion.

Even without integrals, we have derived just about all of the important properties of exponential functions without introducing the limit concept. Something very like a limit does creep in when we can only *estimate* $m_b = \ln b$. But I would argue that this is different from the usual limit definition of derivative. In the first place, it is perfectly legitimate to leave the definition of m_b implicit. We know that it is the slope of a particular line, and we can estimate it to any desired accuracy. We deal in a similar way with $\sqrt{2}$, which can be characterized geometrically as the diagonal of a square of unit side, and which we can only approximate numerically. In the second place, the development above avoids the construction that so vexed Berkeley. He rightly criticized the use of an incremental value that is positive for one step of the construction and zero for another. This inconsistency is eliminated by defining a derivative using the limit concept. But there is no inconsistency to eliminate in the development presented above. So, even if one insists that the definition of the constant m_b ultimately takes the form of a limit, the significance of that limit is almost incidental. We need it to define a particular number, but not to derive the differentiation rule.

11.4 Sine and Cosine

Our development of the rules of differentiation is almost complete. We just need to find the derivative of the sine function. Then, the identity $\cos x = \sin(x + \pi/2)$ gives us the derivative of the cosine, and the remaining trigonometric functions and their inverses can be differentiated using the rules we already have.

Once again we will consider the existence of tangent lines self-evident. In particular, we assume that the graphs of $f(x) = \sin x$ and $g(x) = \cos x$ have tangent lines at $x = 0$, and so are differentiable there. Because of the symmetry of the cosine curve, we observe that the tangent line at $x = 0$ must be horizontal. After all, since reflection across the y-axis leaves the cosine curve unchanged, it leaves the tangent line unchanged. That is possible only if the tangent line is horizontal. Thus, $g'(0) = 0$. For the sine we cannot determine the slope at the origin so easily. For now we will denote $f'(0)$ by m_s, deferring consideration of its numerical value.

The differentiability of f and g at 0 implies the existence of $T_{f,0}$ and $T_{g,0}$ with the following properties. At $x = 0$ both $T_{f,0}$ and $T_{g,0}$ vanish, while for $x \neq 0$,

$$T_{f,0}(x) = \frac{f(x) - f(0)}{x - 0} - m_s = \frac{\sin x}{x} - m_s$$

and

$$T_{g,0}(x) = \frac{g(x) - g(0)}{x - 0} - 0 = \frac{\cos(x) - 1}{x}.$$

Next, we turn our attention to $x = a$. To show that f is differentiable there, we must exhibit the function $T_{f,a}$ and the constant m_f as specified in the definition of differentiability.

As a preliminary step, define $h = x - a$, so that the difference quotient can be written

$$\begin{aligned}
\frac{f(x) - f(a)}{x - a} &= \frac{\sin(x) - \sin(a)}{x - a} \\
&= \frac{\sin(a + h) - \sin(a)}{h} \\
&= \frac{\sin(h)\cos(a) + \sin(a)\cos(h) - \sin(a)}{h} \\
&= \cos(a)\frac{\sin(h)}{h} + \sin(a)\frac{\cos(h) - 1}{h}.
\end{aligned}$$

For $x \neq a$, we observe that $h \neq 0$, so the definitions of $T_{f,0}$ and $T_{g,0}$ imply

$$\frac{f(x) - f(a)}{x - a} = \cos(a)(T_{f,0}(h) + m_s) + \sin(a)T_{g,0}(h).$$

This suggests defining

$$T_{f,a}(x) = \cos(a)T_{f,0}(x - a) + \sin(a)T_{g,0}(x - a).$$

and

$$m_f = \cos(a)m_s$$

for then condition (i) is clearly satisfied. For condition (ii), we compute

$$T_{f,a}(a) = \cos(a)T_{f,0}(0) + \sin(a)T_{g,0}(0) = 0.$$

Thus condition (ii) is also satisfied. In conclusion, $f(x) = \sin x$ is differentiable at $x = a$ and the derivative is given by $f'(a) = m_s \cos(a)$.

Expressed in a slightly different form, we have shown that $(\sin x)' = m_s \cos x$. Using the identity $\cos(x) = \sin(x + \pi/2)$, we derive the corresponding result $(\cos x)' = -m_s \sin x$. At this point we need to know the value of m_s. From the standard limit based development, we know $m_s = 1$ so that the familiar rule $(\sin x)' = \cos x$ follows. But in the context of the current development, we cannot yet justify the conclusion $m_s = 1$.

Here, for the first time, we seem to be stymied. To reach our goal we need to determine the slope of the sine curve at $x = 0$. For the derivative of an exponential like 2^x we need the slope at $x = 0$ as well, but that is $\ln 2$, an irrational number. It makes sense to settle for an approximation (or an implicitly defined symbolic form) in that case. In contrast, the needed slope for the sine function is precisely 1.

I see no way to justify that exact value without using the limit formulation for slope of the tangent line, at least in this one instance. But perhaps someone else will succeed where I have failed. It makes an interesting challenge for the reader. On the other hand, even if we can go no further, it is satisfying to see how far we have come with this limitless formulation of derivatives. We have derived all the standard rules, thus justifying the same algorithmic differentiation methods that are found in any standard calculus course, except that we needed something close to limits to prove the chain rule, and we haven't quite completed the rule for the derivatives of the trigonometric functions. It is possible to go still further, to important results like the first derivative test for max/min problems and the inverse function theorem. That is further than we can go in the present volume, but references to the literature will be given in the next section.

11.5   History, References, and Additional Reading

As stated at the start of the chapter, the development presented here follows directly from the work of Descartes. My source for the history of these ideas is a nice article by Suzuki [157]. It gives an account of the work of Descartes and Hudde, developing methods for tangents and maxima/minima based on the idea of repeated roots of equations. As described by Suzuki, these methods did not follow the approach here, but depended instead on techniques for identifying double roots.

The definition for derivative given in this chapter is essentially the same as a definition due to Carathéodory, as described by Kuhn [103]. It is also closely related to the Fréchet formulation of derivative, which is often used in advanced calculus courses to define the total derivative of a function from $\mathbb{R}^n$ to $\mathbb{R}^m$. The two formulations are considered together in an article by Acosta and Delgado [1]. Together [1] and [103] prove some of the main results of differential calculus, including rules of differentiation, the inverse function theorem, and the first derivative test.

For more on George Berkeley's critique of calculus, see [94, Section 13.5.1].

12

Two Calculusian Miracles

As part of our tour of the Calculusian Republic, it is fitting to spend some time at the very heart of the realm. In this place miracles are to be witnessed. Indeed, there are two miracles that are absolutely central to the subject that is Calculusia. For casual visitor and old hand alike, deep understanding of Calculusia is impossible without fully appreciating them.

Paradoxically, the miracles of which I write are so evident, so much a part of the fabric of daily Calculusian life that they are generally overlooked. It is often thus. Truly miraculous phenomena are frequently taken for granted. Life, love, consciousness, the cosmos: they are a part of everyday experience. They seem utterly mundane when we notice them at all.

Thus it is for the two central miracles of calculus, which may be described thus:

Miracle 1: Much out of little.   Calculus, as a study of real functions, illuminates the fundamental properties of continuity and differentiability, and their powerful consequences. Miraculously, these powerful properties are shared by essentially all members of the broad class of *Elementary Functions*, and this can be proven in an amazingly concise way. Both the breadth of the class of functions to which the proof applies, and the ease with which the proof is propounded are parts of the miracle.

Miracle 2: More accuracy for less effort.   Calculus, as a basis for accurate models of real phenomena, is the limiting case of discrete approximation. Discrete approximations are conceptually simple, intuitively appealing, but evidently of limited accuracy. Passing to the limit requires sophisticated and subtle reasoning, but it produces much more accurate models, and in the idealized world of mathematics, banishes error entirely. It is a miracle that this is possible at all. Even more miraculously, in passing from the discrete models to the more accurate limiting case, the formulation and analysis of the models becomes dramatically easier.

These miracles are thoroughly familiar to all visitors to the Calculusian Republic. But their very familiarity obscures their significance. Tourists rush about at a frenetic pace, anxious to see the sights and experience the marvels of this land. Is it any wonder that they overlook two miracles that appear on every prospect?

The readers of this book are more serious aficionados of mathematical culture. You have made extended tours of mathematical regions, not once, but many times. Together, we have explored some of the outer provinces, following little-known paths to obscure destinations, seeking out vistas new even to the most seasoned traveler. Now we return to the center. Join me in a prolonged meditation on two Calculusian miracles, and in appreciating just how miraculous they are.

12.1 The First Miracle

Elementary calculus is full of theorems that describe the behavior of continuous and differentiable functions. For example, the Intermediate Value Theorem says when a function f is continuous for $a \leq x \leq b$, the range of the function must include every number between $f(a)$ and $f(b)$. Suppose that you want to solve an equation $f(x) = d$, and suppose that you have found a and b for which $f(a) < d$ and $f(b) > d$. Then if f is continuous between a and b, there is a solution to your equation between a and b.

Another theorem is: If f is continuous for $a \leq x \leq b$ then f assumes an absolute maximum and an absolute minimum over the interval. This tells us that certain max/min problems have solutions, just as the Intermediate Value Theorem tells us that certain equations have solutions.

A third example is the First Derivative Test. This states that if a function f is continuous on the closed interval $[a, b]$, differentiable on the open interval (a, b), and has a local maximum at an interior point c, then $f'(c) = 0$. Related results tell us that f is increasing on intervals where the derivative is positive, and decreasing on intervals where the derivative is negative; that the second derivative must be zero at an inflection point; that a function's graph is concave up where its second derivative is positive and concave down where its second derivative is negative. These results are important when interpreting graphs produced by computers and graphing calculators. Such graphs result from plotting a large number of points. In between these points, we assume that the graph is a straight line, or some other simple curve. But how can we be sure this is valid? How do we know that important details might not be found by zooming in on a small part of the graph? Calculus gives us the tools to answer these questions. It can tell us all the points where the graph changes from increasing to decreasing, or from concave up to concave down. Between these points, the graph can only exhibit very tame behavior, much like a straight line.

One more example: the Mean Value Theorem is a statement about functions continuous on a closed interval $[a, b]$ and differentiable on its interior (a, b). It is widely used to prove other results, including those in the preceding paragraph, as well as the Fundamental Theorem of Calculus, Taylor's Theorem, and many others.

There are plenty of other examples of theorems that hold for continuous or differentiable functions. Practically every theorem you find in a calculus text deals with such functions. So, you might expect introductory calculus courses to pay great attention to one question: How do we know whether a function is continuous or not?

But no. Hardly any attention at all is paid to this question. If you ask a typical calculus student which functions are continuous, which are not, and how to tell the difference, the only response you are likely to get is a blank stare. Students are not to be blamed. The question is rendered moot by a marvelous happenstance: essentially all functions we encounter in calculus are continuous and differentiable everywhere, aside from obvious exceptions involving division by zero or some other forbidden operation. That means you can apply all of the results of calculus blindly, without worrying about differentiability or continuity.

There are exceptions of course. The function $f(x) = x^{2/3}$ is continuous on the real line but not differentiable at $x = 0$. For this function, it is easy to concoct examples showing that the conclusions of the Mean Value Theorem or the First Derivative Test fail. Similarly, many students will compute $\int_{-1}^{2} \frac{1}{x^2}\,dx$ using the Fundamental Theorem of Calculus without stopping to consider that the integrand has a point of discontinuity within the interval of integration. When you point out that this assigns a negative value to the integral of a positive function, students are stumped about what has gone wrong.

But the exceptions really prove the point. The fact is, ignoring the hypotheses about differentiability and continuity almost never leads you astray. Teachers have to go to some trouble to find examples of the sort that I have just given, and students recognize this. While the prevailing attitude is not logically justified, it is pragmatically valid: all the functions you meet in calculus are differentiable and continuous wherever they are defined.

The point of these comments is not to condemn students for behaving in this way. Rather, it is to recognize the validity of their belief, and to call attention to how wonderful it is. Moreover, it is not at all difficult to support this belief rigorously. Here is a brief outline, as it appears in most treatments of calculus.

First, there is a definition for *limit*, which permits continuity and differentiability to be defined in turn. Next, appealing to the definition, we see that a few functions are differentiable: x, e^x, $\sin x$, and constants. The definition of differentiability is also used to prove the standard rules for differentiation: the sum and difference rules, product and quotient rules, and the chain rule. It follows that any function definable in terms of x, e^x, $\sin x$, and constants, by way of arithmetic operations, function composition, or function inversion, is differentiable, and differentiability implies continuity.

This is an astounding development. The scope of the conclusion, in comparison to the difficulty of the proof, is staggering. Consider inverse functions alone, for which differentiability appears via the chain rule. This tells us that the curly root function (Chapter 2) and glog (Chapter 9) are differentiable wherever they are defined. Even more impressive, the demonstration is constructive. Not only can we easily conclude that a function *has* a derivative, we can give a simple formula for computing it. Continuity is established almost as an afterthought. This set of ideas goes beyond *astounding*. It surpasses *staggering*. It is nothing short of miraculous.

Miraculous though it is, few visitors to Calculusia pay it much attention. To the novice, there is an inexorable progression of ideas as a semester or year course unfolds, leaving little time for contemplation. Problems are set, and the methods of calculus invariably provide solutions. It seems hardly worth noticing that there *are* solutions and that we *can* find them. They would not be exercises in a textbook were it otherwise. Meanwhile, for old calculus hands, the progression of ideas becomes thoroughly familiar and commonplace — nothing that makes one think of miracles.

Do not be deceived, there is indeed a miracle to behold. The elementary functions are so beautifully behaved, and so unbelievably useful for describing real phenomena. It is little wonder that they have become the focus of study for so great a part of the mathematics curriculum. Calculus tells us exactly why this is so with remarkable concision and clarity.

That brings us to the second Calcusian miracle. To understand it, one must recognize how powerful a tool calculus is for describing real phenomena and predicting how they evolve. We take up these ideas next.

12.2   Why Is Calculus Important?

Calculus is important for many reasons. At every turn in mathematics one finds calculus, or at the very least its shadows and echos. But among those who study calculus, who appreciate its power and apply its methods, mathematicians are a small minority. Calculus is an indispensable tool throughout the natural and social sciences, engineering, and technology. And in these fields, by far the greatest majority of applications involve differential equations.

The applications are not about differential equations, but rather arise when formulating and studying mathematical representations of real phenomena. The mathematical representations are called *models*. The phenomena may be natural, such as weather, biological processes, or the propagation of light or sound. They may also be manmade, like aircraft, automobiles, cell phones, and computers. Some are so completely artificial that even the laws governing their evolution are human inventions, financial markets being a notable example. What they all have in common is change, and the point of a model is to describe and predict such change.

That is done by representing aspects of a phenomenon by variables corresponding to attributes that are quantifiable and measurable, at least in principle. Relationships among the variables are expressed in equations. This restricts the variables to sets of values that are consistent with the equations. If the model is accurate, the variable values describe possible states of the phenomenon under study. Often a model admits a state that has not been observed, and specifies the conditions under which it can arise. That is what we call a prediction.

This description is necessarily vague. It is not my purpose here to elucidate the intricacies of mathematical modeling, nor to survey models in all their variety. Rather, I will make a few general points. First, the overall methodology of mathematical modeling is remarkably effective. Second, the vast majority of models involve differential equations. They are formulated in the language of calculus and analyzed using the techniques of calculus. And third, the great significance of calculus to affairs beyond mathematics rests largely on the first two points: calculus-based mathematical models have been incredibly effective in an astonishing spectrum of applications.

To illustrate just how effective these models can be, we consider an example from astrodynamics. The application of differential equations to model celestial phenomena was one of the first successes of calculus, and continues to be one of its most impressive accomplishments. A case in point is provided by the Mars Rover missions described in Sidebar 12.1. These missions are a nearly unbelievable feat of engineering.

At the start of each mission, a spacecraft is launched into space to begin a 300 million mile journey lasting seven months. For all but a few minutes of the trip, the spacecraft is coasting. It is kicked violently out of Earth orbit, gets a few corrective nudges mid-course, and arrives at Mars with enough fuel to establish an orbit and then land with its payload intact.

Think about what is involved in planning such a mission. When the spacecraft exits Earth orbit, you cannot aim for where Mars is today, but rather for where Mars will be seven months later. You have to predict what path the spacecraft will follow during its seven months of drifting. The margin of acceptable error is infinitesimal. Aiming wrong by the smallest fraction of a degree will put the spacecraft thousands of miles off course after 300 million miles.

The design of the missions accommodates some error, it is true. There are a few mid-course corrections to fine-tune the trajectory. But for these to be productive, you have to know where the spacecraft is, where it is heading, and how it is oriented. As explained in Sidebar 12.1, the instrumentation for the mission determined spacecraft position to an accuracy of one part in 200 million.

I find this nearly unbelievable. Disregarding the little course corrections, you have to project the spacecraft with just the right initial velocity to get to its target. In proportion to the speed involved, it is analogous to a game of intercontinental miniature golf. Your golf ball must go 14,000 miles, more than half way around the world, uphill all the way, and has to hit a moving target. And just to make things interesting, you have to hit the ball while standing on a spinning merry-go-round!

The Rovers both reached their destinations, and were safely landed on the surface of Mars. To this day, they continue to travel the Martian landscape, sending back a wealth of data. These accomplishments reflect the presence of extremely accurate models for celestial motion, as well as powered and unpowered space flight. At the foundation of these models lies Newtonian physics, augmented with extremely detailed and accurate representation of the forces involved, with a healthy dose of feedback and differential correction on the side. Going down yet another level, Newtonian physics stands upon calculus. Thus, the success of the Mars Rover missions is a testimonial to the power of calculus.

This is meant to dramatize the effectiveness of mathematical models. It is but one instance in an immense spectrum of mathematical models across the sciences, engineering, and other fields. The central role of calculus in the formulation and analysis of these models explains in large measure why calculus is so important.

The contribution of calculus can be better appreciated by asking: what would we do without calculus? Had calculus never been invented, what alternatives for analysis would be available? One particularly natural alternative is to use discrete models. Where calculus-based models are expressed in terms of differential equations, the discrete models use *difference* equations. There is a close analogy between differential equations and difference equations. And as we shall see, a differential equation can often be understood as the result of carrying a difference equation to its logical extreme. This is where the second Calculusian miracle occurs. Accordingly, we now turn to a discussion of discrete models and difference equations.

12.3   Discrete Models of Change

In this section we will consider several examples of discrete models. In each we will see how they arise as natural first approximations and how they are expressed using difference equations. The difference equations can be studied using elementary (non-calculus) methods, and in the simplest instances a complete analysis is readily found. In more complicated refinements of the models, the analysis becomes more difficult or impossible. We will also see that adopting a difference equation model for a continuous process has an obvious disadvantage, and how its effects can be reduced.

Navigation on the Way to Mars

In the summer of 2003, NASA launched two space probes bound for Mars. Each spacecraft journeyed 320 million miles over the course of seven months. Fig. 12.1 shows the trajectory of the first of the two missions, launched June 10, carrying the Mars Rover named Spirit. The later mission, launched July 7, carried Spirit's twin, Discovery.

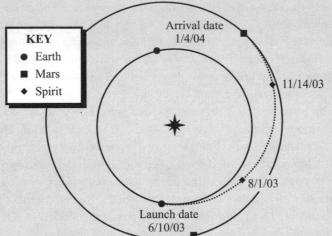

Figure 12.1. Spirit's trajectory to Mars.

Navigating the probes to a soft landing on Mars is a nearly unbelievable feat of engineering. Because they can carry very little fuel, the spacecraft are in free fall for almost the entire journey to Mars, coasting along with no active propulsion. The initial launch sends them on their way, with just a few opportunities to make course corrections included in the flight plan.

Successful execution of the flight plan depends on highly accurate predictions of the trajectories of the spacecraft and of Mars. The course corrections require equally accurate determination of spacecraft location and orientation. If you don't know where you are, or where you are headed, how can you correct your course? And if you don't know how you are oriented, how can you aim your rocket?

A Tank Model. Given a water tank, with water flowing in at the top and out near the bottom, as illustrated in Fig. 12.3, assume that the in- and out-flows are balanced, so the volume in the tank is constant. Suppose that there is some substance dissolved in the water in the tank, but that the water flowing in is pure. We will develop a model for the amount of the dissolved substance present at any time.

For concreteness sake, let us make the following additional assumptions: the volume of the tank is 100 gallons; water flows in and out at a rate of 10 gallons per minute; and the dissolved substance, say salt, initially amounts to 400 grams dissolved in the tank. Let $A(t)$ be the amount of salt in the tank at time t, where A is measured in grams and t in minutes.

Navigation on the Way to Mars (cont.)

This is where navigation and attitude sensors come in. Estimates of spacecraft location and orientation used several different methods. One, radio ranging, determines the spacecraft's distance from tracking stations on the Earth by timing a signal sent to the spacecraft and back. A second, doppler analysis, finds the spacecraft's speed relative to the Earth using radio signal frequency shifts. In addition, the spacecraft uses sensors to analyze its position relative to the sun and other stars, and for attitude determination.

Overall, the determination of spacecraft position is incredibly accurate. At the end of its trajectory, when the spacecraft is near Mars, its position can be determined to within a kilometer. Relative to the distance from Earth, some 200 million kilometers, that works out to 99.9999995% accuracy.

Figure 12.2. Mars Rover Spirit.

Sidebar 12.1 (cont.)

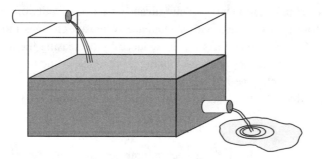

Figure 12.3. A water tank.

In one minute, 10 gallons of water flow out of the tank. That is $1/10$ of the total, and so carries with it $1/10$ of the salt. Thus, when $t = 1$ we expect there to be 360 grams of salt in the tank. In the next minute, $1/10$ of the tank is again drained, this time carrying away 36 grams of salt. So, at time $t = 2$ there will be 324 grams of salt. In each succeeding minute, the amount of salt in the tank is reduced by $1/10$.

This can be expressed by the equation

$$A_{n+1} = .9A_n$$

(called a *difference equation*), where A_n is the value of A after n steps of the model. Here, A_n is the same as $A(t)$ with $t = n$, but that will not always be the case. If we made our computations in half-minute intervals, then n of those intervals would represent $n/2$ minutes, so that $A_n = A(t)$ for $t = n/2$.

The difference equation is *recursive*, it shows how each A_n depends on the preceding value A_{n-1}. We can also determine how each A_n depends on n. Computing the first few stages of the model, we find the following pattern:

$$A(1) = .9(400)$$
$$A(2) = .9^2(400)$$
$$A(3) = .9^3(400)$$
$$\vdots$$
$$A(n) = .9^n(400).$$

In general, after n minutes, the model predicts $A(n) = 400(.9^n)$. This gives A as a function of time, and is called the *solution* of the difference equation. Note that whereas the difference equation arises naturally from our understanding of the system, it is the solution that is most useful in predicting the system's future behavior.

This is a discrete model, because it treats the water tank system as if it changes abruptly at one minute time intervals. We know that this is an approximation. In each minute some mixing will occur. The water flowing out at the start of the minute will carry a greater concentration of salt than the water flowing out at the end of the minute. Thus, the amount of salt leaving the tank in any minute will not be equal to one tenth of the salt present at the start of the minute. In the model we neglect this detail. We also neglect any nonuniformity in the distribution of salt in the tank, and ignore the mechanism by which the pure water mixes with the water in the tank. These effects are overlooked for simplicity, and we assume (or hope) that the simplification will not significantly alter the fidelity of the model.

Modeling a continuous process with a sequence of abrupt changes introduces what is called *discretization error*. Its effects can be reduced by decreasing the length of the time step. We can repeat the earlier analysis looking at changes in the tank at increments of .1 minutes. Then in each time interval the amount of salt in the tank is decreased by a hundredth (rather than a tenth), leading to the difference equation

$$A_{n+1} = .99^n A_n,$$

and its solution $A_n = 400(.99^n)$. Since each minute accounts for 10 steps of the model, an elapsed time of t corresponds to $n = 10t$ steps. Thus, as a function of time, the solution becomes $A(t) = 400(.99^{10t})$. Then, since $.99^{10} \approx .9044$, the revised model gives $A(t) \approx 400(.9044)^t$ as compared to the original model's $A(t) = .9^t$. In similar fashion we can formulate a discrete model for any specified time step. Making the step size δ, we find

$$A(t) = 400(1 - .1\delta)^{t/\delta}.$$

Now we can apply the model with any step size we wish, compare the results for different step sizes, and assess the impact of discretization error.

In this first example, because we considered a simple phenomenon, the discrete approach works quite well. Any improvement from using calculus would be minute, and far short of miraculous. Let us look at some more complicated models.

A Revised Tank Model. As a variation, suppose that salt is added to the tank at the rate of 250 grams per minute. In each minute one tenth of the water in the tank will flow out, and so the amount of salt in the tank will diminish in the same proportion. But now there will also be an addition of 250 grams of salt, so the difference equation is

$$A_{n+1} = .9A_n + 250.$$

We can find a solution by looking at the first several stages of the model. Because at each step we multiply the preceding result by .9 and then add 250, we see

$$A_0 = 400$$
$$A_1 = .9(400) + 250$$
$$A_2 = .9^2(400) + 250(.9) + 250$$
$$A_3 = .9^3(400) + 250(.9^2) + 250(.9) + 250$$
$$\vdots$$
$$A_n = .9^n(400) + 250(.9^{n-1} + .9^{n-2} + \cdots + 1).$$

Although this is more complicated than the first tank model, it is again possible to express the solution as a function of time. To do so requires summing a geometric progression. We use

$$1 + .9 + .9^2 + \cdots + .9^{n-1} = \frac{1 - .9^n}{1 - .9}$$

to derive

$$A_n = 400(.9^n) + 2500(1 - .9^n) = 2500 - 2100(.9^n).$$

This is subject to the same criticism as the first tank model — it is incorrect to hold the amount of salt in the tank constant over each time step. We can again improve the results by decreasing the size of the time step. With step size of δ, we get

$$A_{n+1} = (1 - .1\delta)A_n + 250\delta.$$

The solution is

$$A_n = (1 - .1\delta)^n(400) + 250\delta((1 - .1\delta)^{n-1} + (1 - .1\delta)^{n-2} + \cdots + 1),$$

and hence

$$A_n = 2500 - 2100(1 - .1\delta)^n. \tag{1}$$

Since $A_n = A(t)$ for $t = n\delta$, the solution can also be written

$$A(t) = 2500 - 2100(1 - .1\delta)^{t/\delta}.$$

Incorporating the time step as a parameter in the solution makes it possible to study how changing the time step affects the model, but no matter how small a time step is chosen, there will still be some discretization error. And we were able to find an equation for A_n only by summing a finite series. We will see in the following examples that solutions to difference equations often involve finite series that cannot be evaluated as simply as a geometric series. In these cases, the expression for A_n becomes much less convenient for further analysis.

The modified tank model can be employed in many interesting applications. Reservoirs with reactants flowing in and a mixture flowing out at a constant rate are found in models for chemical reactions, pollution in a body of water, absorption or metabolization of drugs in the human body, the spread of information through various forms of communication, and the propagation and dissipation of heat, among others. Being able to model such systems accurately is not just an idle pursuit.

The Snowplow Problem.   A snowplow can clear snow from the road at the constant rate of 1,000,000 cubic feet per hour. Suppose that snow is falling at the rate of three inches per hour, and that the blade on the plow is 10 feet wide. If the snowplow starts when the snow is a foot deep, how long will it take to travel 100 miles?

We can develop a discrete model for this problem, beginning with a crude analysis. Say the plow travels s miles in the first hour. It clears a volume of snow s miles long, 10 feet wide, and (holding the depth of snow constant over the hour) a foot deep. That amounts to $5280s \times 10 \times 1 = 52,800s$ cubic feet. But we know that the snowplow can clear 1,000,000 cubic feet per hour. Therefore, we must have $52,800s = 1,000,000$, hence $s = 10,000/528$. We take this for s_1, the distance traveled in the first hour (in miles).

In the second hour we assume a depth of 1.25 feet because an additional three inches of snow have fallen. Proceeding as before, we find the distance traveled in the second hour is $s_2 = 10,000/(528 \cdot 1.25)$. In the next hour, with snow 1.5 feet deep, the distance traveled will be $s_2 = 10,000/(528 \cdot 1.5)$.

Generalizing from these results, in hour n, the distance traveled will be

$$s_n = \frac{10,000}{528(1 + .25(n-1))}.$$

The total distance traveled in n hours is thus

$$S_n = \frac{10,000}{528}\left(\frac{1}{1.00} + \frac{1}{1.25} + \frac{1}{1.50} + \cdots + \frac{1}{1 + .25(n-1)}\right),$$

or

$$S_n = \frac{10,000}{528}\sum_{k=0}^{n-1}\frac{1}{1 + .25k}.$$

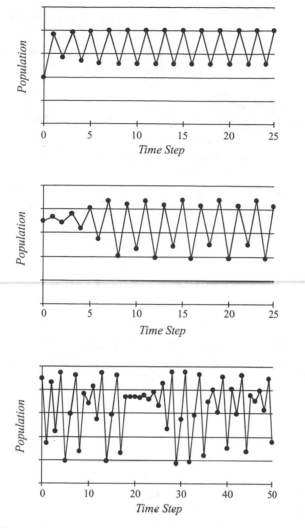

Figure 12.6. Oscillation and chaos in discrete logistic growth.

To see how logistic models differ from others we have seen, let us consider an example. Suppose that the equilibrium population size is 10,000. Then, representing P in units of 10,000 we have $P^* = 1$. For m we choose a value of .5. With $L = P^* + 1/m = 3$, we have

$$P_{n+1} = .5(3 - P_n)P_n.$$

Starting with an initial population size of P_0 how does this model evolve? The first several terms are

$$P_1 = 1.5P_0 - .5P_0^2$$

$$P_2 = \frac{9}{4}P_0 - \frac{15}{8}P_0^2 + \frac{3}{4}P_0^3 - \frac{1}{8}P_0^4$$

$$P_3 = \frac{27}{8}P_0 - \frac{171}{32}P_0^2 + \frac{171}{32}P_0^3 - \frac{465}{128}P_0^4 + \frac{27}{16}P_0^5 - \frac{33}{64}P_0^6 + \frac{3}{32}P_0^7 - \frac{1}{128}P_0^8.$$

We can now attempt to answer the original question: how long will it take to travel 100 miles? Some experimentation shows that $S_9 \approx 96.2$ and $S_{10} \approx 102.03$. Thus, the model predicts it will take between 9 and 10 hours for the snowplow to travel 100 miles.

This first analysis is crude. The problem is that the depth of the snow is changing continuously. It is not valid to assume that the snow remains a foot deep for the entire first hour, and so our computations of both the volume of snow cleared and the distance traveled in the process are incorrect, reflecting once again discretization error. And as before, we proceed next to use a shorter time step.

Let us reduce the time interval to .01 hour. The amount of snow cleared in each time step will then be 10,000 rather than 1,000,000 cubic feet, and in each time step the snow depth will increase by .0025 rather than .25 feet. This leads to

$$s_n = \frac{100}{528(1 + .0025(n - 1))}$$

for the distance traveled in the nth time step, and

$$S_n = \frac{100}{528} \sum_{k=0}^{n-1} \frac{1}{1 + .0025k}$$

for the total distance over n time steps.

We expect the version of the model with a shorter time step to be more accurate than the original one, because the depth of the snow changes much less over each time step. But as before, in order to consider the effects of step size, we should generalize to an arbitrary time step δ. This leads to

$$s_n = \frac{10,000\delta}{528(1 + .25\delta(n - 1))} \tag{2}$$

and

$$S_n = \frac{10,000\delta}{528} \sum_{k=0}^{n-1} \frac{1}{1 + .25\delta k}. \tag{3}$$

This is similar to what we saw in the modified tank model, but here the series cannot be replaced with a simple algebraic expression. Nevertheless, we can use the equation for S_n to solve the snowplow problem for a variety of different δ's, comparing the results to gauge the effect of step size. We also know that no matter how short a time step we use, there is still a discretization error because snow depth is not constant over each time step.

The modified tank model and snowplow problem might appear to differ in one significant way because in the latter we did not formulate a difference equation. However, in the derivation of an expression for S_n we are implicitly using a difference equation, because

$$S_{n+1} = S_n + s_{n+1}. \tag{4}$$

Any model that accumulates incremental changes in a variable can be formulated in a difference equation in this fashion. Our next example has this structure, and is similar to the snowplow problem in other ways, too.

Incident Solar Energy. A solar panel converts sunlight into electricity. If the panel lies fixed, the angle of incidence of the sun's rays changes over the course of a day. This is illustrated in Fig. 12.4, where the angle of incidence is θ. The energy collected by the panel depends on θ. When the sun is directly overhead, with $\theta = \pi/2$, the energy produced by the panel is at a maximum.

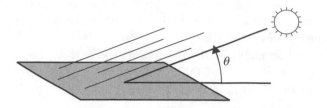

Figure 12.4. Sunlight arrives at a solar collector with incidence angle θ.

To model this, a reasonable first hypothesis is that the energy produced by the solar panel is proportional to $\sin \theta$. Assuming that the panel generates $P = A \sin \theta$ kilowatts, what is the total energy produced in a day?

An accurate model should take into account the way the elevation of the sun changes over a day. This depends on latitude and the day of the year. To simplify, let us assume that θ varies linearly from 0 to π over a period of 12 hours. Thus $\theta = (\pi/12)t$, where t is the time in hours and $t = 0$ at sunrise.

Following a by-now familiar approach, let us subdivide the day into twelve one-hour periods. For each period we hold θ fixed, so we collect $A \sin \theta$ kilowatt-hours of energy. The total energy for the day is then

$$E = A \left(\sin 0 + \sin \frac{\pi}{12} + \sin \frac{2\pi}{12} + \cdots + \sin \frac{11\pi}{12} \right),$$

or

$$E = A \sum_{k=0}^{11} \sin \frac{k\pi}{12}.$$

Using n time steps of length $\delta = 12/n$ hours results in

$$E = A\delta \sum_{k=0}^{n-1} \sin \frac{\pi\delta}{12}k.$$

Logistic Population Growth. As a final example, we examine a discrete model for population growth. As a first approximation, it is commonly assumed that populations grow geometrically, at least at early stages. There is an obvious intuitive appeal to this idea. Since population growth occurs through reproduction, it stands to reason that in any time period the amount of increase will be roughly proportional to the size of the population. This amounts to assuming that

$$P_{n+1} = P_n + \alpha P_n \tag{5}$$

where P_n is the population at the nth time step and α is a constant representing the percentage increase over the time step. Letting $r = 1 + \alpha$,

$$P_{n+1} = rP_n,$$

which is the same difference equation that occured in the first tank model. Its solution is $P_n = P_0 r^n$.

In the first tank model, A is reduced by a fixed fraction α with each time step, so $r = 1 - \alpha < 1$ and A decreases toward a limit of 0. Here $r = 1 + \alpha > 1$ and $P_n = P_0 r^n$ grows exponentially.

For real populations exponential growth is not sustainable because environmental factors impose a limit on population size. To incorporate this in the model we can change $r = 1 + \alpha$ from a constant to a factor that depends on the population size. As the population gets larger, we would like α to decrease. Taking the dependence on population size to be linear leads to what are called *logistic growth* models.

We can be more specific about the variation of α if we imagine that the population size approaches an equilibrium P^*. At that value of P there should be no further growth, so that $\alpha(P^*) = 0$. Thus, we envision $\alpha(P)$ as a linear function decreasing to a root at P^*. Representing the slope as $-m$ we find that $\alpha(P) = m(P^* - P)$. Substituting this in (5) produces

$$P_{n+1} = (1 + m(P^* - P_n))P_n, \tag{6}$$

or

$$P_{n+1} = m(L - P_n)P_n, \tag{7}$$

where $L = P^* + 1/m$. This is a general difference equation for logistic growth. In it the equilibrium population size P^* does not appear directly, but can be expressed as $L - 1/m$.

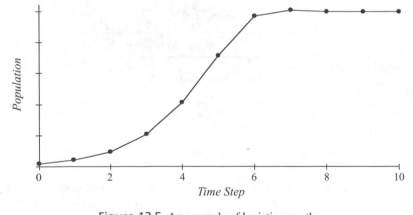

Figure 12.5. An example of logistic growth.

For certain combinations of m and L, discrete logistic growth models behave as we expect. When the population is small its growth is roughly exponential, but as the population grows the rate of growth decreases and the population size levels off. A representative graph is shown in Fig. 12.5. But the methods we used earlier to find solutions to difference equations do not work for logistic growth. Applying the difference equation through several iterations does not produce a simple pattern of results. In fact, for some values of m and L the behavior of the model is very complicated, exhibiting surprising oscillations or even chaotic results with no discernible order. Some examples are shown in Fig. 12.6. It is thus not surprising that logistic growth difference equations do not generally have simple algebraic solutions.

See how the dependence of P_n on P_0 grows increasingly complex as n increases. This contrasts with earlier examples, where simple patterns permitted us to find solutions for the difference equations. Here, no such simple patterns appear.

What about the effect of discretization error? Can we change the logistic growth difference equation to reflect different size time steps?

If in a unit of time the population increases by αP, and if we assume the increase is spread uniformly over the interval, then we expect an increase of $\alpha P/2$ in the first half of the interval and the same amount in the second half. Similarly, if the interval is divided into n parts of length $\delta = 1/n$, we expect the population to grow by $\delta \alpha P$ in each part. In this development, nothing prevents α from depending on P. Thus, if $\alpha = m(P^* - P)$ for a population of size P and one unit of time, then over a subinterval of length δ we replace α with $\delta m(P^* - P)$. This leads to the difference equation

$$P_{n+1} = (1 + m\delta(P^* - P_n))P_n,$$

an analog of (6). For particular values of δ we can explore the behavior of this equation, and that gives some insight into the effect of the time interval. But without any algebraic formulation of the solution P_n, this is hard to generalize.

What Do These Examples Show?   Several general features emerge from these examples. In each case we have a situation we wish to analyze. With no notion of calculus, it would be natural to use discrete models, and they can be formulated readily enough. But we have seen a variety of outcomes. In the first example we found a closed-form solution. In others the solutions were in terms of finite series, while in logistic growth we derived no expression whatever for the solution.

A feature common to all the models is discretization error. Its effects can be lessened by reducing the size of the time step, and in each model we formulated a difference equation in which the step size appears as a parameter. With the exception of the logistic growth difference equation, we expressed the solutions in terms of the step size, making it convenient to assign step sizes of any magnitude we choose. But an analysis of the effects of step size is hindered by the absence of closed-form expressions for the solutions, and in the case of logistic growth, by the absence of any expression at all.

12.4   The Second Miracle

Now we are in a position to appreciate the second Calculusian miracle. In all of the examples, to eliminate discretization error, we would like to reduce the step size to the vanishing point. That is exactly what calculus allows us to do. You might expect there to be a price to pay. Solutions of the difference equations are already fairly complicated, or even nonexistent. To gain essentially infinite accuracy, surely even greater complication is to be expected.

Not so. Passing to the limiting case not only gives better accuracy, it also greatly simplifies the algebraic formulation of the models. Where the original discrete version of the model involves a finite series that cannot be simplified, in the limiting case there is a simple algebraic formula. Even for logistic growth the solution is both explicit and simple.

Does it not seem strange that you can at one stroke make the models both simpler and also more accurate? It is like having your cake and eating it too! This is an incredible stroke of good fortune. And it should have come as a complete surprise. Before anyone figured out how calculus works, there would have been no reason to predict the simplification that it would produce. That it does work out so nicely is simply astonishing. It is wonderful. It is a miracle.

To see it in action, let us revisit the examples.

The Revised Tank Model.   Reducing the step size to zero leads to a differential equation, as follows. Subtract A_n from both sides of

$$A_{n+1} = (1 - .1\delta)A_n + 250\delta$$

to get

$$A_{n+1} - A_n = -.1\delta A_n + 250\delta$$

and hence

$$\frac{A_{n+1} - A_n}{\delta} = -.1A_n + 250.$$

Since the fraction on the left shows the change ΔA in variable A that occurs over time interval $\delta = \Delta t$, the equation can be written

$$\frac{\Delta A}{\Delta t} = -.1A + 250.$$

Taking the limit as $\Delta t \to 0$ gives

$$A'(t) = -.1A(t) + 250,$$

a differential equation. It provides a model for the tank system that is conceptually similar to the discrete model, but on an instantaneous scale. There is no discretization error, because the value of $A(t)$ is not held constant for any time interval. With a differential equation, we permit the dependent variable to change continuously, modeling its impact instantaneously.

There are various ways to solve differential equations. Here is one that works well in the examples we will consider. Rewrite the equation in the Leibniz notation

$$\frac{dA}{dt} = -.1A + 250$$

and then rearrange it to get

$$\frac{dA}{-.1A + 250} = dt.$$

Integrate both sides. On the right, let t go from 0 to an unspecified time, again represented by t. On the left, we choose limits consistently, so the lower limit, corresponding to $t = 0$, is the initial value of A, 400 grams. At the upper limit we will have $A(t)$, the amount of salt in the tank at time t. This leads us to

$$\int_{400}^{A(t)} \frac{dA}{-.1A + 250} = \int_0^t dt.$$

Carrying out the integration,

$$-10\ln(250 - .1A)\Big|_{400}^{A(t)} = t$$

and so
$$\ln(250 - .1A(t)) - \ln 210 = -.1t.$$

Further manipulation leads to
$$\frac{250 - .1A(t)}{210} = e^{-.1t}$$

so
$$A(t) = 2500 - 2100e^{-.1t}.$$

Beautiful. Succinct. This equation is slightly simpler than the corresponding discrete version (1) (from which it can be obtained as a limiting case). More importantly, the method of derivation was simpler, and it eliminates the discretization error.

The Snowplow Problem.   We will again pass from a difference equation to a differential equation, although there is another way to proceed. Because (3) defines S_n, the distance traveled in the first n intervals, as a Riemann sum, letting δ go to zero leads directly to a definite integral. We will not solve the snowplow problem in this way, but it is recommended as an exercise.

To derive a differential equation, we need a difference equation with step size δ. Substitute the expression for s_n in (2) into (4) to obtain
$$S_{n+1} = S_n + \frac{10,000\delta}{528(1 + .25\delta n)}$$

and then rearrange as
$$\frac{S_{n+1} - S_n}{\delta} = \frac{10,000}{528(1 + .25\delta n)}.$$

Let $\Delta t = \delta$ and $t = \delta n$. Then $S_n = S(t)$ and the equation becomes
$$\frac{\Delta S}{\Delta t} = \frac{10,000}{528(1 + .25t)}.$$

As Δt goes to zero, the difference equation becomes the differential equation
$$\frac{dS}{dt} = \frac{10,000}{528(1 + .25t)}.$$

Its solution, $S(t)$, is the distance the snowplow covers by time t.

Since $S = 0$ at $t = 0$, the solution to the differential equation can be obtained from
$$\int_0^{S(t)} dS = \frac{10,000}{528} \int_0^t \frac{dt}{1 + .25t}.$$

Evaluating the integrals and simplifying leads to
$$S(t) = \frac{2500}{33} \ln(1 + t/4). \tag{8}$$

This is beautifully simple when compared with the discrete model analog (3). So, we get both a more faithful model of the problem and greater algebraic simplicity. Moreover, (8) permits us to do something that we could not do with (3) — solve for t. Doing so produces
$$t = 4e^{33S/2500} - 4.$$

Now we can easily answer the original question: How long does it take the snowplow to go 100 miles? The answer is $t = 4e^{33/25} - 4$, or approximately 10.97 hours.

Incident Solar Energy.   For the solar energy model, it is again possible to proceed either by forming a differential equation, or by passing directly from a Riemann sum to a definite integral. It is not necessary to repeat all the details. Instead, let us jump directly to this result:

$$E(t) = A \int_0^t \sin\left(\frac{\pi t}{12}\right) dt,$$

where $E(t)$ is the energy collected by time t. Integration leads to

$$E(t) = \frac{12A}{\pi}\left(1 - \cos\left(\frac{\pi t}{12}\right)\right)$$

with the result that $E(12) = 24A/\pi$. The exact value of the result is not the point of course. Rather, this example shows again how calculus leads us to answers that are both simpler and more accurate.

Logistic Population Growth.   The final example is the most impressive. In the discrete logistic model, we were not able to formulate any sort of solution to the difference equation. For it, of all the discrete models we considered, we found the least useful results. Nevertheless, the continuous version of the logistic model is as easy to analyze as any of the others.

We begin with a difference equation for the discrete logistic model with time step δ,

$$P_{n+1} = (1 + m\delta(P^* - P_n))P_n.$$

Expand the right side of this equation, producing

$$P_{n+1} = P_n + m\delta(P^* - P_n)P_n,$$

so

$$\frac{P_{n+1} - P_n}{\delta} = m(P^* - P_n)P_n,$$

leading to the differential equation

$$\frac{dP}{dt} = m(P^* - P)P.$$

If the population is P_0 at time $t = 0$, integration now gives us

$$\int_{P_0}^{P(t)} \frac{dP}{P(P^* - P)} = m \int_0^t dt.$$

Integrate the right and perform a partial fractions decomposition on the left to arrive at

$$\frac{1}{P^*} \int_{P_0}^{P(t)} \frac{1}{P} + \frac{1}{P^* - P} \, dP = mt$$

and hence

$$\ln\left(\frac{P(t)}{P^* - P(t)}\right) - \ln\left(\frac{P_0}{P^* - P_0}\right) = P^* mt.$$

Solving for $P(t)$ produces

$$P(t) = \frac{P^* P_0}{P_0 + (P^* - P_0)e^{-P^* mt}}.$$

This is a general form for continuous logistic growth. It is well known, and has a pleasing graph (Fig. 12.7) and many interesting properties. To get it, we had to work a bit harder than in the earlier examples. But the benefit derived is worth the effort (and anyway we can use a computer algebra system for the integration and manipulation if we wish).

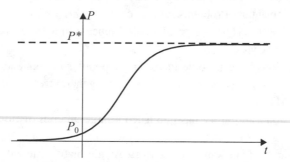

Figure 12.7. A continuous logistic growth curve.

The continuous model has obvious advantages over the discrete analog. First and foremost, it is possible to express P as an explicit function of t only in the continuous case. Then, too, the continuous model is well behaved, with none of the chaos that can arise in the discrete model. This suggests that the chaotic behaviors are artifacts of discretization error.

The logistic model highlights the second Calculusian miracle. What appears initially as a plausible discrete model for limited growth leads to difficulties. We end up unable to solve the difference equation that arises, as propagation of the model reveals an arithmetic complexity that increases frightfully. In some situations, the future behavior of the model grows surprisingly complicated, even approaching apparent randomness. Under these circumstances, it is not obvious that allowing the step size to decrease to zero will improve the situation, and there would be no reason to expect an algebraic simplification. Surprisingly, both of these occur. Using a differential equation rather than a difference equation gives us a result that is at once both more plausible and algebraically simpler. This is marvelous. It is fabulous good fortune. It is a miracle.

12.5   History, References, and Additional Reading

There are many textbooks devoted separately to each of the three topics, difference equations, differential equations, and mathematical modeling. My book [78] offers a gentle introduction to discrete models, covering difference equations, logistic growth, and the appearance of chaos. The modified tank model is referred to there as a mixed model. Although the pace of the book is slow and the assumed mathematical experience of the audience is modest, the presentation describes several applications of discrete models, and may be of interest on that account.

At a higher level, [150] uses modeling as a vehicle for teaching calculus. It discusses both the continuous and discrete logistic growth models, and includes a historical sketch describing the work of Belgian mathematician P. F. Verhulst in the nineteenth century. Verhulst is often credited with originating the logistic growth differential equation as a model for limited growth. Continuous logistic growth is also frequently discussed in calculus texts. Some representative examples are [17, 57]. The first presents a nice assortment of applications from different areas of study.

The snowplow problem is a variant of one that has appeared in a number of differential equation texts. Recent papers concerning it are [109, 169]. An early version of the problem appeared in the Elementary Problems section of the *American Mathematical Monthly* in 1937 [12]. According to [169], the problem also appears in a 1942 differential equations textbook by R. P. Agnew.

Difference equations have become increasingly popular in connection with dynamical systems and chaos. For a discussion see [141]. Some general treatments of chaos are provided by [38, 39, 61].

The Jet Propulsion Laboratory has published a large amount of information about the Mars rover missions on the internet. See [116] to learn more about navigation, [117] for trajectory details, and [118] for more on sensors. An alternative site with information about the rover missions is [119].

References

Links to all internet references are provided at the webpage for this book:
http://www.maa.org/ume.

[1] Ernesto Acosta G. and Cesar Delgado G. Fréchet vs. Carathéodory, *American Mathematical Monthly*, vol. 101, no. 4 (1994), pp. 332–338.

[2] Roger C. Alperin. A mathematical theory of origami constructions and numbers, *New York Journal of Mathematics*, vol. 6 (2000), pp. 119–133, http://nyjm.albany.edu:8000/j/2000/6-8.html.

[3] A. R. Amir-Moéz. Khayyam's solution of cubic equations, *Mathematics Magazine*, vol. 35, no. 5 (1962), pp. 269–271.

[4] Branden Archer and Eric W. Weisstein. Lagrange interpolating polynomial, *MathWorld–A Wolfram Web Resource* (webpage), http://mathworld.wolfram.com/LagrangeInterpolatingPolynomial.html, 2005.

[5] G. C. Archibald and Richard G. Lipsey. *An Introduction to Mathematical Economics: Methods and Applications,* Harper and Row, New York, 1976.

[6] Author Unknown. Solution to Calculus Problem 356, proposed by F. B. Finkel, *American Mathematical Monthly*, vol. 22, no. 1 (1915), pp. 24–26

[7] David Bange and Linda Host. Does a parabola have an asymptote?, *College Mathematics Journal*, vol. 24, no. 4 (1993), pp. 331–342.

[8] E. J. Barbeau. *Polynomials.* Springer-Verlag, New York, 1989.

[9] Ed Barbeau. Fallacies, flaws, and flimflam, *College Mathematics Journal*, vol. 22, no. 3 (1991), pp. 220–223.

[10] William Barnier and Douglas Martin. Unifying a family of extrema problems, *College Mathematics Journal*, vol. 28, no. 5 (1997), pp. 388–391.

[11] Roger R. Bate, Donald D. Mueller, and Jerry E. White. *Fundamentals of Astrodynamics,* Dover, 1971.

[12] J. A. Benner. *Problem E275, American Mathematical Monthly*, vol. 44, no. 4 (1937), p. 245.

[13] George Berkeley. *The Analyst,* in James Newman, *The World of Mathematics,* vol. 1, Simon and Schuster, NY, 1956.

[14] Elwyn R. Berlekamp. *Algebraic Coding Theory*, McGraw-Hill, New York, 1968.

[15] William P. Berlinghoff and Fernando Q. Gouvêa. *Math Through the Ages — a Gentle History for Teachers and Others*, Oxton House and the Mathematical Association of America, Washington, DC, 2004.

[16] Leah Wrenn Berman. Folding beauties, *College Mathematics Journal*, vol. 37 , no. 3 (2006), pp. 176–186.

[17] Geoffrey C. Berresford and Andrew M. Rocket. *Brief Applied Calculus*, 2nd ed., Houghton Mifflin, NY, 2000.

[18] W. H. Bixby. Discussions: Graphical Solution of Numerical Equations, *American Mathematical Monthly*, vol. 29, no. 9 (1922), pp. 344–346.

[19] Viktor Blåsjö. The isoperimetric problem, *American Mathematical Monthly*, vol. 112, no. 6. (2005), pp. 526–566.

[20] R. P. Boas, Jr. Partial sums of infinite series, and how they grow, *American Mathematical Monthly*, vol. 84, no. 4 (1977), pp. 237-258.

[21] Folkmar Bornemann and Stan Wagon. A perplexing polynomial puzzle, revisited, *College Mathematics Journal*, vol. 36, no. 4 (2005), p. 288.

[22] Jonathan M. Borwein and Robert Corless. Emerging tools for experimental mathematics, *American Mathematical Monthly*, vol. 106, no. 10 (1999), pp. 889-909.

[23] Peter Borwein and Tamás Erdélyi. *Polynomials and Polynomial Inequalities*, Springer-Verlag, New York, 1995.

[24] Raymond T. Boute. Moving a rectangle around a corner - geometrically, *American Mathematical Monthly*, vol. 111, no. 5 (2004), pp. 435–437.

[25] Carl B. Boyer. *A History of Mathematics,* 2nd. ed., Wiley, New York, 1968.

[26] Phillips V. Bradford. *Visualizing solutions to n-th degree algebraic equations using right-angle geometric paths: Extending Lill's Method of 1867* (webpage), http://www.concentric.net/~Pvb/ALG/rightpaths.html.

[27] Manuel Bronstein, Robert M. Corless, James H. Davenport, and D.J. Jeffrey. Algebraic properties of the Lambert W function from a result of Rosenlicht and of Liouville, *Integral Transforms and Special Functions*, vol. 19, no. 10 (2008), pp. 709–712.

[28] Roger Chalkley. Cardan's formulas and biquadratic equations, *Mathematics Magazine*, vol. 47, no. 1 (1974), pp. 8–14.

[29] Alpha C. Chiang. *Fundamental Methods of Mathematical Economics,* 3rd ed., McGraw-Hill, New York, 1984.

[30] Mircea I. Cîrnu. Sums of integer powers of the roots of a polynomial, preprint, 2008.

[31] R. J. Clarke. Sequences of polygons, *Mathematics Magazine*, vol. 52, no. 2. (1979), pp. 102–105.

[32] C. J. Coe. Problems on maxima and minima, *American Mathematical Monthly*, vol. 49, no. 1, (1942), pp. 33–37.

[33] John Cooper. *Ladder Problem Query*, private correspondence, 2006.

[34] R. M. Corless, G. H. Gonnet, D. E. G. Hare, D. J. Jeffrey, and D. E. Knuth. On the Lambert *W* function, *Advances in Computational Mathematics*, vol. 5, no. 4 (1996), pp. 329–359.

[35] Richard Courant. *Differential and Integral Calculus, Volume 2,* translated by E. J. McShane, Interscience, New York, 1949.

[36] L. M. Court. Envelopes of plane curves, *American Mathematical Monthly*, vol. 57, no. 3 (1950), pp. 168–169.

[37] H. T. Croft, K. J. Falconer, and R. K. Guy. *Unsolved Problems in Geometry,* Springer-Verlag, New York, 1994.

[38] Robert L. Devaney. *An Introduction to Chaotic Dynamical Systems,* Addison-Wesley, Redwood City, CA, 1987.

[39] ——. *Chaos, Fractals, and Dynamics; Computer Experiments in Mathematics*, Addison-Wesley, Menlo Park, CA, 1990.

[40] L. E. Dickson. Solution to problem 219, *American Mathematical Monthly*, vol. 11, no. 4 (1904), p. 93.

[41] Henry Ernest Dudeney. *Amusements in Mathematics*, T. Nelson and sons, ltd, London, 1917.

[42] William Dunham. *Euler, The Master of Us All*, Mathematical Association of America, Washington, DC, 1999.

[43] J. E. Eaton. The fundamental theorem of algebra, *American Mathematical Monthly*, vol. 67, no. 6. (1960), pp. 578–579.

[44] William Holding Echols. *An Elementary Text-Book on the Differential and Integral Calculus,* Holt, New York, 1902.

[45] The Economics Professor. *Envelope Theorem* (webpage), Arts & Sciences Network, http://www.economyprofessor.com/economictheories/envelope-theorem.php .

[46] B. Carter Edwards and Jerry Shurman. Folding quartic roots, *Mathematics Magazine*, vol. 74, no.1 (2001), pp. 19–25.

[47] Harold M. Edwards. *Galois Theory*, Springer-Verlag, New York, 1984.

[48] Howard Eves. A note on envelopes, *American Mathematical Monthly*, vol. 51 , no. 6 (1944), p. 344.

[49] H. E. Fettis. On various methods of solving cubic equations, *National Mathematics Magazine*, vol. 17, no. 3 (1942), pp. 117–130.

[50] S. R. Finch. Moving sofa constant, section 8.12 in *Mathematical Constants,* Cambridge University Press, Cambridge, England, pp. 519–523, 2003.

[51] T. J. Fletcher. Easy ways of going round the bend, *Mathematical Gazette* vol. 57 (1973), pp. 16–22.

[52] James H. Foster and Jean J. Pedersen. On the reflective property of ellipses, *American Mathematical Monthly*, vol. 87, no. 4 (1980), pp. 294–297.

[53] J. S. Frame. Machines for solving algebraic equations, *Mathematical Tables and Other Aids to Computation*, vol. 1, no. 9. (1945), pp. 337-353.

[54] C. G. Fraser. Isoperimetric problems in the variational calculus of Euler and Lagrange, *Historia Mathematica*, vol. 19, no. 1 (1992), pp. 4–23.

[55] Orrin Frink, Jr. A method for solving the cubic, *American Mathematical Monthly*, vol. 32, no. 3 (1925), p. 134.

[56] Peter J. Giblin. Zigzags, *Mathematics Magazine*, vol. 74, no.4 (2001), pp. 259–271.

[57] Larry J. Goldstein, David C. Lay, and David I. Schneider. *Calculus and Its Applications,* 10th ed., Pearson, Upper Saddle River, NJ, 2004.

[58] Gene E. Golub and Charles F. Van Loan. *Matrix Computations,* Johns Hopkins, Baltimore, 1983.

[59] Harry Gonshor. Remarks on asymptotes, *Mathematics Magazine*, vol. 41, no. 4 (1968), pp. 197–198.

[60] J. W. Green. On the envelope of curves given in parametric form, *American Mathematical Monthly*, vol. 59, no. 9 (1952), pp. 626–628.

[61] Denny Gulick. *Encounters with Chaos*, McGraw, New York, 1992.

[62] James Haddon. *Examples and Solutions of the Differential Calculus,* Virtue, London, 1862.

[63] William J. Hazard. *Algebra Notes*, Vantage Press, 1952, p. 187.

[64] Catherine Hassell and Elmer Rees. The index of a constrained critical point *American Mathematical Monthly*, vol. 100, no. 8 (1993), pp. 772–778.

[65] Morton J. Hellman. A unifying technique for the solution of the quadratic, cubic, and quartic, *American Mathematical Monthly*, vol. 65, no. 4 (1959), pp. 274–276.

[66] Magnus R. Hestenes. *Optimization Theory: the Finite Dimensional Case,* Wiley, New York, 1975.

[67] Michael Hoy, John Livernois, Chris McKenna, Ray Rees, and Thanasis Stengos. *Mathematics for Economics,* 2nd ed., MIT Press, Cambridge, MA, 2001.

[68] Deborah Hughes-Hallett, Andrew M. Gleason, William G. McCallum, et al. *Calculus Single and Multivariable,* 2nd ed., Wiley, New York, 1998.

[69] Thomas Hull. *Project Origami: Activities for Exploring Mathematics*, A K Peters, Wellesley, MA, 2006.

[70] ———. A note on "impossible" paper folding, *American Mathematical Monthly*, vol. 103, no. 3 (1996), 242–243.

[71] ———. *Origami Mathematics* (webpage), http://www.merrimack.edu/~thull/origamimath.html.

[72] John Hymers. *A Treatise on the Theory of Algebraic Equations,* 2nd ed., Deighton and Rivington, Cambridge, England, 1811, pp 41-43, 77–79.

[73] Mark Kac. Can one hear the shape of a drum?, *American Mathematical Monthly*, vol. 73, no. 4, part 2: Papers in analysis (1966), 1–23.

[74] ———. *Enigmas of Chance: an Autobiography*, Harper & Row, New York, 1985.

[75] Dan Kalman. The generalized Vandermonde matrix, *Mathematics Magazine*, vol. 57, no. 1 (1984), pp. 15–21.

[76] ———. The maximum and minimum of two numbers using the quadratic formula, College Mathematics Journal, vol. 15, no. 4 (1984), pp. 329–330.

[77] ———. Geometrically asymptotic curves, College Mathematics Journal, vol. 16, no. 3 (1985), pp. 199–206.

[78] ———. *Elementary Mathematical Models, Order Aplenty and a Glimpse of Chaos*, Mathematical Association of America, Washington, DC, 1997.

[79] ———. A matrix proof of Newton's identities, *Mathematics Magazine*, vol. 73, no. 4 (2000), pp 313–315.

[80] Dan Kalman. Suen's method for solving cubics — explained, *AMATYC Review*, vol. 22, no. 1 (2000), pp. 15–21.

[81] ——. A generalized logarithm for exponential-linear equations, *College Mathematics Journal*, vol. 32, no. 1 (2001), pp. 2–14.

[82] ——. Integrability and the glog function, *AMATYC Review*, vol. 23, no. 1 (2001), pp. 43–50.

[83] ——. James E. White 1946-2004, *FOCUS*, vol. 24, no. 8 (2004), p. 33. Also online at http://www.maa.org/pubs/nov04.pdf.

[84] ——. Virtual empirical investigation: concept formation and theory justification, *American Mathematical Monthly*, vol. 112, no. 9 (2005), pp 786–798.

[85] ——. The maximal deflection on an ellipse, *College Mathematics Journal*, vol. 37, no. 4 (2006), pp. 250–260.

[86] ——. Solving the ladder problem on the back of an envelope, *Mathematics Magazine*, vol. 80, no. 3 (2007), pp 163–182.

[87] ——. Uncommon mathematical excursions: supplementary resources (webpage), http://www.maa.org/ume, 2008.

[88] ——. An elementary proof of Marden's theorem, *American Mathematical Monthly*, vol. 115, no. 4 (2008), pp 330–338.

[89] ——. The Most Marvelous Theorem in Mathematics, *Journal of Online Mathematics and Its Applications* vol. 8, article ID 1663, http://mathdl.maa.org/mathDL/4/?pa=content&sa=viewDocument&nodeId=1663, 2008.

[90] Dan Kalman and Warren Page. Nested polynomials and efficient exponentiation algorithms for calculators, *College Mathematics Journal*, vol. 16, no. 1 (1985), pp. 57–60.

[91] Dan Kalman and James White. A simple solution of the cubic, *College Mathematics Journal*, vol. 29, no. 2 (1998), pp. 415–418.

[92] ——. Polynomial equations and circulant matrices, *American Mathematical Monthly*, vol. 108, no. 9 (2001), pp 821–841.

[93] Toni Kasper. Integration in finite terms: the Liouville theory, *Mathematics Magazine*, vol. 53, no. 4 (1980), pp 195–201.

[94] Victor J. Katz. *A History of Mathematics — an Introduction*, HarperCollins, New York, 1993.

[95] I. B. Keene. A perplexing polynomial puzzle, *College Mathematics Journal*, vol. 36, no. 2 (2005), pp. 100, 159.

[96] R. Bruce King. *Beyond the Quartic Equation,* Birkhäuser, Boston, 1996.

[97] Murray S. Klamkin and R. G. McLenaghan. An ellipse inequality, *Mathematics Magazine*, vol. 50, no. 5 (1977), pp. 261–263.

[98] Felix Klein. Elementarmathematik vom höheren Standpunkte aus, 2nd. ed, Teubner, Leipzig, 1911; Springer, Berlin, 1926; p. 267.

[99] Israel Kleiner. Thinking the unthinkable: the story of complex numbers (with a moral), *Mathematics Teacher*, vol. 81, no. 3 (1988), pp. 583–92.

[100] Morris Kline. *Advanced Calculus,* 2nd ed., Wiley, New York, 1972.

[101] ——. *Mathematical Thought from Ancient to Modern Times,* Oxford University Press, New York, 1972.

[102] Donald Knuth. *The Art of Computer Programming,* Volume 2: Seminumerical Algorithms, 3rd ed., Addison-Wesley, Reading, MA, 1997.

[103] Stephen Kuhn. The derivative à la Carathéodory, *American Mathematical Monthly*, vol. 98, no. 1 (1991), pp. 40–44.

[104] J. L. Lagrange. *Analytical Mechanics,* translated from the *Mécanique analytique,* nouvelle édition of 1811 by Auguste Boissonnade and Victor N. Vagliente, Kluwer, Dordrecht, 1997.

[105] ——. *Théorie des fonctions analytiques contenant les principes du calcul différentiel, dégagés de toute considération d'infiniment petits ou d'évanouissans, de limites ou de fluxions, et réduits à l'analyse algébrique des quantités finies,* De l'imprimerie de la République, Paris, 1797.

[106] *Lambert W function* (webpage), http://en.wikipedia.org/wiki/Lambert_W_function.

[107] Roland E. Larson, Robert P. Hostetler, and Bruce H. Edwards. *Calculus with Analytic Geometry,* sixth edition, Houghton Mifflin, Boston, 1998.

[108] Leon M. Lederman and Christopher T. Hill. *Symmetry and the Beautiful Universe,* Prometheus Books, Amherst, NY, 2004.

[109] Jerome L. Lewis. The meeting of the plows: a simulation, *College Mathematics Journal*, vol. 26, no. 5 (1995), pp. 395–400.

[110] H. E. Licks. *Recreations in Mathematics,* D. Van Nostrand, New York, 1917, p. 89.

[111] Lill Eduard, Techniker und Genie-offizier, in *Österreichisches Biographisches Lexikon: 1815–1950,* vol. 5, Österreichische Akademie der Wissenschaften, Vienna, pp 214–215.

[112] Brian J. Loe and Nathanial Beagley. The coffee cup caustic for calculus students, *College Mathematics Journal*, vol. 28, no. 4 (1997), pp. 277–284.

[113] D. Steven Mackey, Niloufer Mackey, Christian Mehl and Volker Mehrmann. Structured polynomial eigenvalue problems: good vibrations from good linearizations, *SIAM Journal of Matrix Analysis and Applications*, vol. 28, Issue 4 (2006), pp. 1029–1051.

[114] ——. Palindromic polynomial eigenvalue problems: good vibrations from good linearizations, Technical Report, DFG Research Center, Technische Universität, Berlin, Germany. April 2005.

[115] Elena Anne Marchisotto and Gholam-Ali Zakeri. An invitation to integration in finite terms, *College Mathematics Journal*, vol. 25, no. 4 (1994), pp. 295–308.

[116] *Mars exploration rover mission: communications with Earth* (webpage), http://marsrovers.jpl.nasa.gov/mission/comm_nav.html.

[117] *Mars exploration rover mission: mission timeline: cruise* (webpage), http://marsrovers.jpl.nasa.gov/mission/tl_cruise.html.

[118] *Mars exploration rover mission: spacecraft: cruise configuration* (webpage), http://marsrovers.jpl.nasa.gov/mission/spacecraft_cruise.html.

[119] *Mars exploration rovers: mission overview* (webpage), http://www.mars.tv/mer/overview.html.

[120] E. J. McShane. The Lagrange multiplier rule, *American Mathematical Monthly*, vol. 80, no. 8 (1973), pp. 922–925.

[121] G. A. Miller. Geometric solution of the quadratic equation, *Mathematical Gazette*, vol. 12, no. 179, (1925), pp. 500–501.

[122] Christopher Moretti. Moving a couch around a corner, *College Mathematics Journal*, vol. 33, no. 3 (2002), pp. 196–201.

[123] H. A. Nogrady. A new method for the solution of cubic equations, *American Mathematical Monthly*, vol. 44, no. 1 (1937), pp. 36–38.

[124] Newton's identities, *Wikipedia* (webpage), http://en.wikipedia.org/wiki/Newton_identities, 2008.

[125] Jeffrey Nunemacher. Asymptotes, cubic curves, and the projective plane, *Mathematics Magazine*, vol. 72, no. 3 (1999), pp. 183–192.

[126] J. J. O'Connor and E. F. Robertson. *Mark Kac* (webpage), http://www-groups.dcs.st-and.ac.uk/~history/Biographies/Kac.html.

[127] E. J. Oglesby. Note on the algebraic solution of the cubic, *American Mathematical Monthly*, vol. 30, no. 6 (1923), pp. 321–323.

[128] Thomas J. Osler. Cardan polynomials and the reduction of radicals, *Mathematics Magazine*, vol. 74, no. 1 (2001), pp. 26–32.

[129] Ivars Peterson. The Galois story (webpage), *Ivars Peterson's Math Treks*, MAA (1999), http://www.maa.org/mathland/mathtrek_3_1_99.html.

[130] John T. Pettit. A speedy solution of the cubic, *Mathematics Magazine*, vol. 21, no. 2 (1947), pp. 94–98.

[131] B. H. Pourciau. Modern multiplier rules, *American Mathematical Monthly*, vol. 87, no. 67 (1980), pp. 443–452.

[132] Victor V. Prasolov. *Polynomials*, Springer-Verlag, New York, 2004.

[133] C. B. Read. What is an asymptote?, *American Mathematical Monthly*, vol. 46, no. 8 (1939), pp. 498–499.

[134] M. Riaz. Geometric solutions of algebraic equations, *American Mathematical Monthly*, vol. 69, no. 7 (1962), pp. 654–658.

[135] H. L. Rietz. Note on the definition of an asymptote, *American Mathematical Monthly*, vol. 19, no. 5 (1912), pp. 89–90.

[136] Tony Rothman. Genius and biographers: the fictionalization of Evariste Galois, *American Mathematical Monthly*, vol. 89, no. 2 (1982), pp. 84-106. Also available at http://www.physics.princeton.edu/~trothman/galois.html.

[137] Ranjan Roy. Mark Kac and the Cubic, preprint, 2007.

[138] John W. Rutter, *Geometry of Curves,* Chapman & Hall / CRC, Boca Raton, 2000.

[139] Donald G. Saari and John B. Urenko. Newton's method, circle maps, and chaotic motion, *American Mathematical Monthly*, vol. 91, no. 1. (1984), pp. 3–17.

[140] A. Pen-Tung Sah. A uniform method of solving cubics and quartics, *American Mathematical Monthly*, vol. 52, no. 4 (1945), pp. 202–206.

[141] James T. Sandefur. *Discrete Dynamical Modeling*, Oxford, NY, 1993.

[142] Charles W. Schelin. Calculator function approximation, *American Mathematical Monthly*, vol. 90, no. 5 (1983), pp. 317–325.

[143] Mark Schwartz. The chair, the area rug, and the astroid, *College Mathematics Journal*, vol. 26, no. 3 (1995), pp. 229–231.

[144] Angelo Segalla and Saleem Watson. The flip-side of a Lagrange multiplier problem, *College Mathematics Journal*, vol. 36, no. 3 (2005), pp. 232–235.

[145] Junpei Sekino. n-ellipses and the minimum distance sum problem, *American Mathematical Monthly*, vol. 106, no. 3 (1999), pp. 193–202.

[146] Paul Shutler. Constrained critical points, *American Mathematical Monthly*, vol. 102, no. 1 (1995), pp. 49–52.

[147] Andrew Simoson. An envelope for a spirograph, *College Mathematics Journal*, vol. 28, no. 2 (1997), pp. 134–139.

[148] ———. The trochoid as a tack in a bungee cord, *Mathematics Magazine*, vol. 73 , no. 3 (2000), pp. 171–184.

[149] David Singmaster. *Sources in Recreational Mathematics, an Annotated Bibliography* (webpage), http://us.share.geocities.com/mathrecsources/.

[150] David A. Smith and Lawrence C. Moore. *Calculus Modeling and Application*, Heath, Lexington, MA, 1996.

[151] M. R. Spiegel. Reciprocal quadratic equations, *American Mathematical Monthly*, vol. 59, no. 3. (1952), pp. 175–177.

[152] D. Spring. On the second derivative test for constrained local extrema, *American Mathematical Monthly*, vol. 92, no. 9 (1985), pp. 631–643.

[153] Ian Stewart. *Why Beauty is Truth, A History of Symmetry*, Basic Books, New York, 2007.

[154] James Stewart. *Calculus Early Transcendentals,* 3rd edition, Brooks/Cole, Pacific Grove, CA, 1995.

[155] Gilbert Strang. Duality in the classroom, *American Mathematical Monthly*, vol. 91, no. 4 (1984), pp. 250–254.

[156] Robert Y. Suen. Roots of cubics via determinants, *College Mathematics Journal*, vol. 25, no. 2 (1994), pp. 115–117.

[157] Jeff Suzuki. The lost calculus (1637–1670): tangency and optimization without limits, *Mathematics Magazine*, vol. 78, no. 5 (2005), pp. 339–353.

[158] Angus E. Taylor and W. Robert Mann. *Calculus: an Intuitive and Physical Approach,* 2nd ed., Wiley, New York, 1977.

[159] George B. Thomas, Jr., Maurice D. Weir, Joel D. Hass, and Frank R. Giordano. *Thomas' Calculus Early Transcendentals,* tenth edition, Addison-Wesley, Boston, 2001.

[160] Abraham A. Ungar. A unified approach for solving quadratic, cubic, and quartic equations by radicals, *Int. J. Comp. Math. Appl.*, vol. 19 (1990), pp. 33–39.

[161] B. L. van der Waerden. *Algebra,* Volume 1, Frederick Ungar, New York, 1970.

[162] Paulo Viana and Paula Murgel Veloso. Galois theory of reciprocal polynomials, *American Mathematical Monthly*, vol. 109, no. 5 (2002), pp. 466–471.

[163] Jas. A. Ward. Using the hessian to solve a cubic equation, *Mathematics Magazine*, vol. 9, no. 8 (1935), pp. 235–240.

[164] William C. Waterhouse. Do symmetric problems have symmetric solutions?, *American Mathematical Monthly*, vol. 90, no. 6 (1983), pp. 378–387.

[165] Thayer Watkins. *The Envelope Theorem and Its Proof* (webpage), San Jose State University, http://www2.sjsu.edu/faculty/watkins/envelopetheo.htm .

[166] Eric W. Weisstein. *Lambert W function* (webpage), From MathWorld—A Wolfram Web Resource, http://mathworld.wolfram.com/LambertW-Function.html.

[167] ——. *Moving Sofa Problem* (webpage), From MathWorld—A Wolfram Web Resource, http://mathworld.wolfram.com/MovingSofaProblem.html.

[168] C. R. White. Definitive solutions of general quartic and cubic equations, *American Mathematical Monthly*, vol. 69, no. 4 (1962), pp. 285–287.

[169] Xiao-peng Xu. The snowplow problem revisited, *College Mathematics Journal*, vol. 22, no. 2 (1991), pp. 139.

[170] George A. Yanosik. A graphical solution for the complex roots of a cubic, *Mathematics Magazine*, vol. 10, no. 4 (1936), pp. 139–140.

[171] Robert C. Yates. *Curves and their Properties,* National Council of Teachers of Mathematics, Reston, VA, 1974.

[172] Doron Zeilberger. A combinatorial proof of Newton's identities, *Discrete Mathematics*, vol. 49, no. 3 (1984), p. 319.

Index

About the Author

Dan Kalman has been writing about and teaching mathematics for 30 years. A graduate of Harvey Mudd College (BS, 1974) and the University of Wisconsin (PhD, 1980) he is a Professor of Mathematics at American University, Washington, DC. He previously held faculty positions at the University of Wisconsin, Green Bay, and Augustana College, Sioux Falls, among other institutions, and worked for several years as an applied mathematician at the Aerospace Corporation. He also served for one year as an Associate Executive Director of the MAA. Kalman's mathematical writing has been recognized with multiple MAA awards: Allendoerfer Awards in 1998 and 2002, Pólya Awards in 1994 and 2002, and an Evans Award in 1997. He is the author of one previous book, *Elementary Mathematical Models*, published by the MAA in 1997. Kalman has served on the Editorial Boards for several MAA publications, including *Mathematics Magazine*, *FOCUS*, *Math Horizons*, and the Spectrum and Classroom Resource Materials book series.